Tech Lab Workbook
Construction and Building Technology

by

E. Keith Blankenbaker
Associate Professor Emeritus
Industrial Technology Education
The Ohio State University, Columbus, Ohio

Publisher
The Goodheart-Willcox Company, Inc.
Tinley Park, Illinois
www.g-w.com

Copyright © 2013
by
The Goodheart-Willcox Company, Inc.

All rights reserved. No part of this work may be reproduced, stored, or
transmitted in any form or by any electronic or mechanical means, including
information storage and retrieval systems, without the prior written permission of
The Goodheart-Willcox Company, Inc.

Manufactured in the United States of America.

ISBN 978-1-60525-812-6

1 2 3 4 5 6 7 8 9 – 13 – 17 16 15 14 13 12

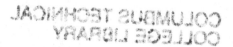

The Goodheart-Willcox Company, Inc. Brand Disclaimer: Brand names, company names, and illustrations for products and services included in this text are provided for educational purposes only and do not represent or imply endorsement or recommendation by the author or the publisher.

The Goodheart-Willcox Company, Inc. Safety Notice: The reader is expressly advised to carefully read, understand, and apply all safety precautions and warnings described in this book or that might also be indicated in undertaking the activities and exercises described herein to minimize risk of personal injury or injury to others. Common sense and good judgment should also be exercised and applied to help avoid all potential hazards. The reader should always refer to the appropriate manufacturer's technical information, directions, and recommendations; then proceed with care to follow specific equipment operating instructions. The reader should understand these notices and cautions are not exhaustive.

The publisher makes no warranty or representation whatsoever, either expressed or implied, including but not limited to equipment, procedures, and applications described or referred to herein, their quality, performance, merchantability, or fitness for a particular purpose. The publisher assumes no responsibility for any changes, errors, or omissions in this book. The publisher specifically disclaims any liability whatsoever, including any direct, indirect, incidental, consequential, special, or exemplary damages resulting, in whole or in part, from the reader's use or reliance upon the information, instructions, procedures, warnings, cautions, applications, or other matter contained in this book. The publisher assumes no responsibility for the activities of the reader.

The Goodheart-Willcox Company, Inc. Internet Disclaimer: The Internet resources and listings in this Goodheart-Willcox Publisher product are provided solely as a convenience to you. These resources and listings were reviewed at the time of publication to provide you with accurate, safe, and appropriate information. Goodheart-Willcox Publisher has no control over the referenced websites and, due to the dynamic nature of the Internet, is not responsible or liable for the content, products, or performance of links to other websites or resources. Goodheart-Willcox Publisher makes no representation, either expressed or implied, regarding the content of these websites, and such references do not constitute an endorsement or recommendation of the information or content presented. It is your responsibility to take all protective measures to guard against inappropriate content, viruses, or other destructive elements.

Table of Contents

Section 1—Introduction to Construction Technology

Chapter 1
What Is Construction Technology? 9
    Chapter 1 Review—What Is Construction Technology? 9
    Activity 1-1—What Is Green Construction? 13
    Activity 1-2—Sidewalk Superintendent 17

Chapter 2
Planning for and Controlling Construction 19
    Chapter 2 Review—Planning for and Controlling Construction. 19
    Activity 2-1—Applying Community Development 21

Chapter 3
Construction Safety 25
    Chapter 3 Review—Construction Safety 25
    Activity 3-1—General Safety. 29
    Activity 3-2—Lab Safety 31

Chapter 4
The Construction Process 35
    Chapter 4 Review—The Construction Process. 35
    Activity 4-1—Comparing Contracts. 39

Chapter 5
Construction Tools and Equipment 43
    Chapter 5 Review—Construction Tools and Equipment 43
    Activity 5-1—Identifying Construction Equipment. 45
    Activity 5-2—Tool Ownership. 49

Section 2—Construction Materials

Chapter 6
Concrete 51
    Chapter 6 Review—Concrete 51
    Activity 6-1—Concrete Tools. 55

Chapter 7
Metals 57
Chapter 7 Review—Metals. 57
Activity 7-1—Metal Materials 61
Activity 7-2—Simple Life-Cycle Cost Analysis 63

Chapter 8
Wood and Wood Products 65
Chapter 8 Review—Wood and Wood Products 65
Activity 8-1—Wood Materials 69

Chapter 9
Masonry, Glass, and Plastics 71
Chapter 9 Review—Masonry, Glass, and Plastics 71
Activity 9-1—Masonry, Glass, and Plastic Identification 75

Section 3—Designing the Project

Chapter 10
Architectural Design 77
Chapter 10 Review—Architectural Design. 77
Activity 10-1—Reading Residential Drawings. 81
Activity 10-2—Green Certification Programs 89
Activity 10-3—Designing a Storage Building, Part 1 91
Activity 10-4—Designing a Storage Building, Part 2 101
Activity 10-5—Designing a Storage Building, Part 3 111
Activity 10-6—Designing a Storage Building, Part 4 115

Chapter 11
Construction Engineering 119
Chapter 11 Review—Construction Engineering. 119
Activity 11-1—Analyzing Commercial Architectural Drawings. 121

Chapter 12
Construction Documentation 125
Chapter 12 Review—Construction Documentation. 125

Section 4—Construction Management

Chapter 13
Project Management 127
Chapter 13 Review—Project Management. 127
Activity 13-1—Responsibilities of Contractors and Employees 129
Activity 13-2—Green Construction Practices 133

Chapter 14
Construction Estimating and Bidding 135
Chapter 14 Review—Construction Estimating and Bidding. 135

Chapter 15
Construction Scheduling 139
Chapter 15 Review—Construction Scheduling 139
Activity 15-1—The Critical Path Method 141

Section 5—Beginning Construction

Chapter 16
Site Preparation 143
Chapter 16 Review—Site Preparation. 143
Activity 16-1—Locating Property Boundaries 145
Activity 16-2—Locating Structures 153

Chapter 17
Earthwork 157
Chapter 17 Review—Earthwork. 157
Activity 17-1—Selecting Earthmoving Equipment 159

Chapter 18
Foundations 161
Chapter 18 Review—Foundations. 161
Activity 18-1—Green Foundation Materials 165
Activity 18-2—Concrete Activity 167
Activity 18-3—Concrete Activity Evaluation. 173

Section 6—Building the Superstructure

Chapter 19
Floors 175
Chapter 19 Review—Floors 175
Activity 19-1—Green Framing Materials 176
Activity 19-2—Floor Joist Layout 181
Activity 19-3—Drawing Floor Frames 183
Activity 19-4—Observation Form—Portable Electric Circular Saw 187

Chapter 20
Walls 191
Chapter 20 Review—Walls. 191
Activity 20-1—Drawing a Wall Frame 193
Activity 20-2—Carpentry Activity 197
Activity 20-3—Carpentry Activity Evaluation 201
Activity 20-4—Observation Form—Miter Saw 203

Chapter 21
Roof and Ceiling Framing 205
Chapter 21 Review—Roof and Ceiling Framing. 205
Activity 21–1—Rafter Layout 209
Activity 21-2—Observation Form—Radial Arm Saw 213

Chapter 22
Enclosing the Structure   215
Chapter 22 Review—Enclosing the Structure 215
Activity 22-1—Masonry Activity 219
Activity 22-2—Masonry Activity Evaluation. 225

Section 7—Installing Plumbing, HVAC, and Communication Systems

Chapter 23
Plumbing Systems   227
Chapter 23 Review—Plumbing Systems 227
Activity 23-1—Green Plumbing Products 233
Activity 23-2—Plumbing Activity. 235
Activity 23-3—Plumbing Activity Evaluation 239

Chapter 24
Heating, Ventilating, and Air-Conditioning (HVAC) Systems   241
Chapter 24 Review—Heating, Ventilating, and Air-Conditioning (HVAC) Systems 241
Activity 24-1—Green HVAC Products 245

Chapter 25
Electrical Power Systems   247
Chapter 25 Review—Electrical Power Systems 247
Activity 25-1—Electrical Activity 253
Activity 25-2—Electrical Activity Evaluation 257

Chapter 26
Communication Systems   259
Chapter 26 Review—Communication Systems 259
Activity 26-1—Installing Communication Products. 263
Construction and Building Technology Contractor's License 267

Section 8—Completing the Project

Chapter 27
Insulating Structures   269
Chapter 27 Review—Insulating Structures. 269
Activity 27-1—Installing Insulation. 273

Chapter 28
Finishing the Building   275
Chapter 28 Review—Finishing the Building. 275
Activity 28-1—Installing Drywall 279
Activity 28-2—Painting. 281

Chapter 29
Landscaping 285
Chapter 29 Review—Landscaping 285
Activity 29-1—Landscaping—Planting. 287
Activity 29-2—Sustainable Landscaping 289

Chapter 30
Final Inspection, Contract Closing, and Project Transfer 291
Chapter 30 Review—Final Inspection, Contract Closing, and Project Transfer 291
Activity 30-1—Transferring the Project. 293

Section 9—Servicing Construction Projects

Chapter 31
Project Operation, Maintenance, and Repair 301
Chapter 31 Review—Project Operation, Maintenance, and Repair 301
Activity 31-1—Project Operation, Maintenance, and Repair (OMR) Calendar. 303
Activity 31-2—Drywall Repair. 305

Chapter 32
Remodeling Buildings 307
Chapter 32 Review—Remodeling Buildings. 307
Activity 32-1—Room Remodeling Project 309

Section 10—Commercial, Industrial, and Engineered Construction

Chapter 33
Dam Construction 319
Chapter 33 Review—Dam Construction 319
Activity 33-1—Commercial, Industrial, and Engineered Construction 323
Activity 33-2—World's Largest Dams. 327

Chapter 34
Bridge Construction 329
Chapter 34 Review—Bridge Construction 329

Chapter 35
Road Construction 333
Chapter 35 Review—Road Construction 333

Chapter 36
Skyscraper Construction 337
Chapter 36 Review—Skyscraper Construction 337

Chapter 37
Pipeline Construction 341
Chapter 37 Review—Pipeline Construction 341

Section 11—Construction and Your Future

Chapter 38
Careers in Construction   343
Chapter 38 Review—Careers in Construction 343
Activity 38-1—Drafting Your Résumé 347
Activity 38-2—Refining Your Résumé 351

Chapter 39
Construction in the Future   353
Chapter 39 Review—Construction in the Future 353
Activity 39-1—Construction and Your Future 355

Construction Company Simulation   357
Module A Floor Plan 359
Module A Elevation 361
Module A Section 363
Module A Plumbing Plan 365
Module A Specifications 367
Module B Floor Plan 369
Module B Elevation 371
Module B Section 373
Module B Plumbing Plan. 375
Module B Specifications 377
Module C Floor Plan 379
Module C Elevation. 381
Module C Section 383
Module C Plumbing Plan 385
Module C Specifications 387
Estimating Work Sheet 389
Job Categories and Standard Data. 393
Materials Price List 395
Bid Blanks 397
Contract 401
Bar Chart Scheduling Form 403
Daily Time Card 405
Building Permit Application 407
Purchase Order 409
Store Accounting Form 417
Accounting System 419
Entering Data in Spreadsheets. 421
Individual Time Summary 423
Checks 425
Individual Earnings Summary. 429
Contract Summary 431

Chapter 1 Review
What Is Construction Technology?

Name _____ Date _____ Class _____

Study Chapter 1 of the text, then answer the following questions.

1. Define *technology*.
   _____
   _____

2. Define *science*.
   _____
   _____

3. How do computers relate to science and technology?
   _____
   _____
   _____

_____ 4. True or False? The discovery of scientific knowledge often precedes the development of technological knowledge.

_____ 5. Using the _____ method, hypotheses are tested through observation and experimentation.

_____ 6. The _____ method solves problems by evaluating solutions, selecting the most appropriate solution, and producing a product to solve the problem.

7. Identify each missing element of the universal systems model shown in the flow chart.

   A. _____
   B. _____
   C. _____
   D. _____

   Inputs → **Processes** → **Outputs**

   Inputs: People, Capital, Knowledge, Materials, Energy, Time, Finance
   Processes: Production, Management
   Outputs: End Results, Products

   Feedbacks

Identify each of the following elements as an input (I), process (P), or output (O).

_____ 8. Design

_____ 9. Desired products

_____ 10. Capital

_____ 11. Time

_____ 12. Organization

_____ 13. Undesirable products

_____ 14. Money

_____ 15. True or False? The types of processes in the universal systems model are production processes and management processes.

_____ 16. The basic construction _____ can be applied to a variety of construction projects.

_____ 17. Information about system performance is called _____.

18. List the seven major technological systems used to satisfy the wants and needs of people.

_____
_____
_____
_____
_____
_____
_____

19. Describe one positive impact of construction technology.

_____
_____
_____

20. Describe one undesirable impact of construction technology.

_____
_____
_____

_____ 21. Since the 1900s, many changes to construction processes have occurred in area of _____.
   A. housing
   B. water treatment
   C. energy and power
   D. All of the above.

Name _____

_____ 22. The construction of large factory buildings is known as _____ construction.

23. Advances have been made in three areas of construction. Name these areas.

_____

_____

_____

_____ 24. True or False? Building codes are enacted to control the use of land.

Name _____ Date _____ Class _____

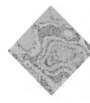

Activity 1-1
What Is Green Construction?

Green Construction features can be found in many chapters of *Construction and Building Technology*. The features focus on different aspects of the green construction movement and introduce many of its concepts. The green construction field continues to grow and evolve. Up-to-the-minute information and trends in green construction can be found by searching the Internet. The ability to find and evaluate information is an important skill.

In this activity, you will learn how to access information on green construction and how to evaluate the information you find. With billions of documents available, the quality of documents varies widely. Not all information is of equal value. The Google search engine ranks pages found in a search. The highest ranked page is based on the number of other pages that include links to this page and the ranking of those pages. This system helps evaluate the quality of the sources, but is not a guarantee that the materials are accurate or unbiased.

There are many search engines to choose from when searching the Internet. Google is commonly used for general searches. The following directions are for use with the Google search engine, but the steps are basically the same for other search engines.

1. Take your seat at a computer. Open the Google home page and type *green construction* in the search box. You may need to press return to begin the search.

2. Near the top of the first page of results and to the right of the word *Search,* you will find the number of results the search produced and the amount of time it took Google to identify these pages. Record these numbers in the blanks.

    Results _____ Time _____

3. Scan the first page of results. Select one of the sources and record the address.

    _____

4. Click on the address to go to the website. Review the first page to find as much of the following information as possible.

    _____

    A. What organization, company, or person is responsible for the website?

    _____

    B. When was the information published?

    _____

    C. On a separate sheet of paper, outline the information that is relevant to your search.

Copyright by The Goodheart-Willcox Co., Inc.

5. Rate the information from 1 (low) to 5 (high) using the following criteria. Record your rating in the first row of Table A.

   Accuracy—Is the information consistent with information on other websites? Is there a way to contact the author? You may delay entering this rating until you have reviewed additional sites.

   Source—What is the domain extension? In general, .edu indicates a site that is registered to a university. Sites registered to federal, state, or local governmental bodies are designated by .gov. These two extensions suggest that the information on the site may be more reliable than those bearing other domain names such as .com or .org.

   Objectivity—Is information accurate and objectively presented? Is advertising limited?

   Currency—Is the publication date given? Is the page up to date?

   Access—Can information be viewed without fees or special browser technology or software?

Table A—Rating Websites						
Name of Website	Accuracy	Source	Objectivity	Currency	Access	Total

6. Return to the first page of results, scroll to the bottom of the page, and click on the number *10*, to take you to the tenth page of results. Select one source and record the address.

   _____

7. Click on the address to go to the website. Review the first page to find as much of the following information as possible.

   _____

   A. What organization, company, or person is responsible for the website?

   _____

   B. When was the information published? _____

   C. On a separate sheet of paper, outline the information that is relevant to your search.

   D. Enter your rating of the information on this site in the second row of Table A.

8. Scroll to the bottom of the page. *Searches related to green construction* lists other topics related to your search. Select one and record the number of results found and the time it took for the search.

   Results _____ Time _____

9. Scan the first page of results. Select one source from this page and record the address.

   _____

Name _____

10. Click on the address to go to the website. Review the first page to find as much of the following information as possible.

    _____

    A. What organization, company, or person is responsible for the website?

    _____

    B. When was the information published? _____

    C. Outline the information that is relevant to your search on a separate sheet of paper.

    D. Enter your rating of the information on this site in Table A.

11. The following distinct topics are part of the broader topic of green construction. Focus your search on one of these topics by entering it in the search box. Circle the topic you selected.

    _____

    - Sustainable construction
    - Life-cycle cost analysis
    - Leadership in Energy and Environmental Design
    - Certified green materials
    - Certified sustainable materials
    - Indoor air quality

12. Record the number of results found and the time it took for the search.

    Results _____ Time _____

13. Scan the first page of results. Select one of the sources and record the address.

    _____

14. Click on the address to go to the website. Review the first page to find the following information.

    _____

    A. What organization, company, or person is responsible for the website?

    _____

    B. When was the information published? _____

    C. On a separate sheet of paper, outline the information that is relevant to your search.

    D. Enter your rating of the information on this site in Table A.

15. Review your outlines from each website. Complete or modify your ratings of the accuracy of the information.

16. On a separate sheet of paper, summarize the information to provide an answer to the original question: What is green construction?

Name _____ Date _____ Class _____

Activity 1-2
Sidewalk Superintendent

In this activity, you will observe construction work being done in your community. You will begin with overall observation of a construction site and move on to more detailed observations. You will also prepare periodic reports on progress being made at the site. Writing a report will help you focus on details and relate the activity to what you are studying in this class.

Each student should locate a different site. If this is not practical, you may observe the same site but focus on different parts of the project. Observe from public property, such as the sidewalk. Do not enter the construction site. In most areas this is considered trespassing and could result in arrest. If anyone from the site asks what you are doing, explain that you are preparing a report for your construction class.

First Report

Locate one to three sites that are easily observed. Record the following for each site:

- Project location—address or nearest intersection.
- Type of construction—pipeline, bridge, office building, home, etc.
- Type of work underway—foundation excavation, window installation, roadbed preparation, etc.

After your instructor reviews the reports from your class, you will receive a specific assignment.

Second Report

Focus on the layout of the site. Sketch the site and the area immediately surrounding it. Your sketch should include:

- Site boundaries.
- Labels indicating what is immediately adjacent to the site, such as street name, types of building, woods or farmland, commercial development, etc.
- Location of driveways or other access from the nearest street.
- Location of materials and equipment and tool storage, if present.
- Names of contractors, developers, architects, and designers involved in the project. Look for signs or vehicles that display company names. If you have the opportunity to talk with someone working on the site, explain what you are doing and ask them to identify the construction companies working on the project.

Subsequent Reports

Focus on a detail and prepare a sketch of that detail. For example, you might see:

- Concrete footer forms in place or being built. Make a sketch showing how the forms are assembled.
- Floor frame being built. Sketch showing how the floor joists are installed around a stairwell opening.
- Steel frame being erected. Sketch a joint between a column and a girder.
- Roof being framed. Sketch one or more of the trusses.

Your sketch may be a section, elevation, plan, or isometric view. Draw the parts to scale as much as possible by estimating their relative size. Label key parts of your sketch and add notes as needed. If you have questions, include them with your sketches.

Finally, list the tools and equipment you saw in use. If you do not know the name of an item, sketch it or write a brief description of it.

Chapter 2 Review
Planning for and Controlling Construction

Name _____ Date _____ Class _____

Study Chapter 2 of the text, then answer the following questions.

_____ 1. True or False? Many older US cities started where they did to meet a need.

_____ 2. An economically justifiable construction project must _____.
   A. satisfy a demand
   B. satisfy a need or provide a service
   C. be well located and buildable
   D. All of the above.

_____ 3. A(n) _____ is a detailed plan that describes how community development goals will be achieved.

4. Explain the difference between a planning board and a city planner.

_____
_____
_____
_____
_____
_____

5. List four types of construction projects that are usually included in a community development plan.

_____
_____
_____
_____

_____ 6. A(n) _____ has a fixed number of access points.
   A. service street
   B. arterial street
   C. limited-access highway
   D. divided highway

7. The transportation segment of a community plan can include what kinds of developments?
_____
_____

_____ 8. Construction of gas, electricity, communication, water, and sewer services is known as _____ construction.

_____ 9. Which of the following is *not* a public service building?
   A. library
   B. grocery store
   C. fire station
   D. hospital

10. What is a trade area?
_____
_____

_____ 11. A poorly maintained structure is considered _____.
   A. deteriorated
   B. enhanced
   C. prime
   D. attractive

12. What typically occurs to community development when there is an economic downturn?
_____
_____

_____ 13. Value can be lost through construction _____.

Name _____ Date _____ Class _____

Activity 2-1
Applying Community Development

The purpose of this two-part activity is to relate community development to the area where you live. Part 1 consists of questions that you answer outside of class. During a subsequent class meeting, you, your classmates, and instructors will discuss the questions and your answers to the Part 1 questions. Questions in Part 2 are answered following the class discussion.

Part 1

Answer the following questions outside of class.

1. Describe the trade area of the community where you live or near where you live if you live outside the city limits.

   _____
   _____
   _____
   _____

2. Identify one example of primary construction in your community.

   _____
   _____
   _____
   _____

3. Is your community growing, stable, or declining?

   _____

4. Name three of the largest employers in your community.

   _____
   _____
   _____

5. Do you know of a new employer that is planning to move or start a business in your community in the near future? If yes, name the new employer.

   _____

Copyright by The Goodheart-Willcox Co., Inc.

6. Do you know of current employers in your community who are planning to hire additional employees in the near future? If yes, list two.

_____

_____

_____

7. Identify two consequences that might occur in your community if the largest employer was to go out of business or move to another community more than 200 miles away.

_____

_____

_____

_____

_____

Part 2 _____

Answer the following questions after your class has discussed their answers to the questions in Part 1.

1. Modify your answer for Question 1 in Part 1 to better describe the trade area of your community.

_____

_____

_____

_____

2. Following the class discussion, would you make any additions to your answer to Question 2 in Part 1?

_____

_____

_____

3. Have you changed your position regarding Question 3 in Part 1? If yes, what is your opinion now? Why did you change your mind?

_____

_____

_____

_____

4. Did you learn of any other large employers in your community? If yes, list up to three *not* listed in your answer for Question 4, Part 1.

_____

_____

_____

Name _____

5. Add two additional new employers who are moving or starting new businesses in your community in the near future.

   _____

   _____

6. Name two existing employers adding employees to those listed in your answer for Question 6, Part 1.

   _____

   _____

7. Add two more consequences that might occur to your community if the major employer goes out of business or moves more than 200 miles away.

   _____

   _____

   _____

   _____

Chapter 3 Review
Construction Safety

Name _____ Date _____ Class _____

Study Chapter 3 of the text, then answer the following questions.

_____ 1. To ensure the safety of constructed products, _____ are enacted and enforced.
   A. laws
   B. codes
   C. product standards
   D. All of the above.

_____ 2. True or False? Codes are usually based on product standards created by manufacturers and oversight organizations in a particular field.

3. What is the purpose of a model code?
   _____
   _____

Choose the code that applies to the installation of each of the following materials and enter the corresponding letter in the blank to the left.

_____ 4. Electrical cable   A. Building code
_____ 5. Water supply pipe   B. Plumbing code
_____ 6. Size of furnace duct   C. Mechanical code
_____ 7. Floor joist   D. Electrical code

8. Why is it important to know about potential side effects of any prescription medicine you are taking?
   _____
   _____

9. Among other requirements, the _____ requires that places of public accommodation be accessible to all individuals.
   _____

_____ 10. Product standards _____.
   A. are nationally recognized minimum requirements for products.
   B. provide a common basis for understanding and comparing products.
   C. are usually developed voluntarily by producers, distributors, or consumers of a product.
   D. All of the above.

_____ 11. True or False? Architects and engineers are responsible for all aspects of project design, excluding safety.

_____ 12. Which of the following statements about job safety is *not* true?
   A. There are few safety hazards on a job site.
   B. Inattentiveness can result in injuries and property damage.
   C. Safety on the job is your personal responsibility.
   D. Shortcuts and speed often contribute to accidents.

13. What is the purpose of OSHA?

_____

_____

14. Identify and briefly describe three items of clothing worn by a carpenter who works framing houses.

_____

_____

_____

_____

_____

15. Identify three protective devices used by a carpenter who works framing houses.

_____

_____

_____

_____ 16. Personal protective equipment includes _____.
   A. eye protection
   B. respirators
   C. hearing protection
   D. All of the above.

_____ 17. A device that buckles around each leg, the waist, and both shoulders is a(n) _____.

Name _____

Match the type of ladder to the correct usage designated for that type.

_____ 18. Type I

_____ 19. Type IA

_____ 20. Type IAA

_____ 21. Type II

_____ 22. Type III

A. Special duty grade made of aluminum or fiberglass.

B. Extra-heavy industrial or professional grade for workloads up to 300 pounds.

C. Commercial grade intended for tasks such as painting.

D. Heavy-duty industrial grade for workloads up to 250 pounds.

E. Household grade intended for use around the home.

23. List five safety precautions to follow when using power tools.

_____
_____
_____
_____
_____
_____
_____

_____ 24. True or False? Unlike power tools, hand tools require no special safety precautions for use.

Match the classification of fire to the combustible material it involves.

_____ 25. Class A

_____ 26. Class B

_____ 27. Class C

_____ 28. Class D

A. Flammable liquids

B. Combustible metals

C. Electrical equipment

D. Ordinary combustibles

29. What is the name of the product shown in the following Material Safety Data Sheet?
_____

```
                           CPVC SOLVENT CEMENT
    SECTION 1              IDENTITY OF MATERIAL
    Trade Name:            OATEY CPVC SOLVENT CEMENT
    Product Numbers:       31127, 31128, 31129, 31130, 31131
    Formula:               CPVC Resin in Solvent Solution
    Synonyms:              CPVC Plastic Pipe Cement
    Firm Name &            OATEY CO.  4700 West 160th Street   P.O. Box 35906  Cleveland,
    Mailing Address:       Ohio  44135, U.S.A.           http://www.oatey.com
    Oatey Phone Number:    (216) 267-7100
    Emergency Phone        For Emergency First Aid call 1-303-623-5716 COLLECT. For
    Numbers:               chemical transportation emergencies ONLY, call Chemtrec at
                           1-800-424-9300
```

Name _____ Date _____ Class _____

Activity 3-1
General Safety

Learning safe work habits and skills is a fundamental goal of this class. Most of the construction work you do will be done in a lab setting. The safety habits and skills you learn in the lab are the same as those used on a construction work site. As the class progresses, you will learn additional safety practices and skills. You will watch demonstrations and practice the techniques shown. Your experiences will improve your skills in each of these areas.

The following discussion points introduce the fundamentals of lab safety. They call special attention to those skills you need to learn and practice now. Safety is everyone's responsibility, every day!

A respectful attitude about safety is very important to your performance. Believing that bad things will not happen to you may result in injury to yourself or others. But being afraid and tentative might also result in injury. The goal is for you to develop an attitude that:

- Respects the importance of safety.
- Encourages you to learn how to safely perform tasks.
- Results in improved safety for everyone in the lab.

Your responsibilities in the lab also include behaving in a way that prevents injuries to you and your classmates. Examples of behavior that might result in injury to yourself or others include:

- Running and horseplay.
- Throwing tools or materials.
- Distracting others by making sudden movements or noises.
- Leaving anything on the floor that could cause a slip or fall.
- Carrying or moving objects in a careless way.

Learning what to do and how to do it safely builds confidence and improves performance. To further improve your skills and performance, study outside of class, participate in demonstrations and discussions, and ask questions when you are not certain about a procedure. Think about safety before you act. Most accidents are avoidable.

Copyright by The Goodheart-Willcox Co., Inc.

Name _____ Date _____ Class _____

Activity 3-2
Lab Safety

In this activity, you will apply the general safety practices to the construction lab. Based on the information discussed in Chapter 3 and Activity 3-1, complete this worksheet on safety topics.

1. Describe three actions that demonstrate a positive attitude toward safety.

   _____
   _____
   _____
   _____
   _____
   _____

2. Describe three actions that demonstrate a negative attitude toward safety.

   _____
   _____
   _____

3. Describe three types of clothing that are appropriate for the lab.

   _____
   _____
   _____

4. Describe three types of clothing that are inappropriate for the construction lab.

   _____
   _____
   _____

5. When do you need to wear eye protection in the lab?

   _____
   _____

6. When do you need to use ear protection in the lab?

   _____
   _____

Copyright by The Goodheart-Willcox Co., Inc.

7. What should you do if you are injured?
   _____
   _____

8. What should you do if the injury is a minor, such as a small cut or splinter?
   _____
   _____

9. List three actions that create a safety hazard for your classmates.
   _____
   _____
   _____

10. Long hair can get caught in revolving tools and machines such as drill, saws, and routers. How can this type of accident be prevented?
    _____
    _____
    _____
    _____

11. Describe various safety practices related to each of the following items.

    Long or loose hair
    _____
    _____

    Hard hats
    _____
    _____

    Eye protection
    _____
    _____

    Loose clothing
    _____
    _____

Name _____

Apron or lab coat

_____
_____
_____

Shoes

_____
_____
_____

Jewelry

_____
_____
_____

Chapter 4 Review
The Construction Process

Name _____ Date _____ Class _____

Study Chapter 4 of the text, then answer the following questions.

_____ 1. Identifying the need or want for a construction project and guiding it through the initial planning process is known as _____.

_____ 2. True or False? Publicly owned projects are initiated by individuals or small groups.

Match each term with the correct definition.

_____ 3. Cost/benefit analysis

_____ 4. Power of eminent domain

_____ 5. Feasibility study

A. A right of the government to take private land for public benefit.

B. An examination of the probable expenses incurred and value gained from a proposed project.

C. An inquiry that determines if a proposed project is practical.

6. Detailed design work involves five steps. Place the five steps in the order they are typically completed:

    _____ Establish budget

    _____ Prepare detailed drawing and specifications

    _____ Reconsider site characteristics

    _____ Review requirement

    _____ Review preliminary plans

7. Briefly describe the responsibilities of a general contractor.

    _____

    _____

8. Briefly describe the responsibilities of subcontractors.
   _____
   _____

9. Name five types of negotiated contracts.
   _____
   _____
   _____
   _____
   _____

_____ 10. In competitive bidding, a contractor is selected based on _____.
   A. location
   B. connections
   C. dates
   D. price

_____ 11. A(n) _____ is an insurance policy that guarantees to the owner that the contractor will complete a project according to the terms of the contract.
   A. performance bond
   B. bid bond
   C. bid document
   D. notice to proceed

12. List eight tasks that are done to prepare a building site for construction.
   _____
   _____
   _____
   _____
   _____
   _____
   _____
   _____

Name _____

13. The typical building construction process follows a sequence. Number the steps in the order in which they are usually completed.

    _____ Finish the building exterior

    _____ Install landscaping

    _____ Frame and enclose the structure

    _____ Lay the foundation

    _____ Finish the building interior

    _____ Install the mechanical system

Match each item with the correct definition.

_____ 14. Warranty

_____ 15. Claim

_____ 16. Punch list

_____ 17. Release of lien

_____ 18. Certificate of completion

A. A document given when payment has been made to settle a mechanic's lien.

B. A request made by a contractor for additional money when it is necessary to do more work than was indicated in the original contract.

C. A document that guarantees the integrity of a job, product, or material.

D. Itemized inventory of all items to be changed, repaired, or replaced in order to complete a project.

E. Document issued by building officials confirming that a building meets all code requirements.

Activity 4-1
Comparing Contracts

In this activity, you will work in small groups. Each group will represent a construction manager. As the construction manager, you are working with a potential client who wants to build a new home. The home is to be large, unique, and in a remote location. The home's design involves challenging work and the client seems uncertain about some features of the design. Your group is to select the type of contract it believes will be attractive to the client and also allow the company to make a profit.

Table A summarizes important characteristics of the various types of contracts.

Table A—Contract Advantages and Disadvantages		
Type of Contract	**Advantages**	**Disadvantages**
Fixed-price	Owner–knows final cost. Contractor–makes more if cost is less.	Owner–plan changes can be expensive. Contractor–pays for cost overruns.
Cost-plus-fixed-fee	Owner–more flexibility to change plan. Contractor–collects fee regardless of cost.	Owner–final cost is unknown until project is completed. Contractor–profit reduced if overhead increases.
Cost-plus-percentage-of-cost	Owner–can make as many changes as desired. Contractor–gets a percentage of any additions to the cost.	Owner–may spend much more than originally budgeted. Contractor–may spend much more time on the project than planned.
Incentive-Share overrun	Owner–encourages contractor to control cost. Contractor–reduces loss from cost overrun.	Owner–less flexibility than cost-plus contracts. Contractor–less opportunity for profit than fixed-price contract.
Incentive-Share savings	Owner–encourages contractor to save money. Contractor–receives half of savings.	Owner–little opportunity to make changes that add to cost. Contractor–limited opportunity for additional profit.

Copyright by The Goodheart-Willcox Co., Inc.

The estimate your company has prepared indicates that the house can be built for $675,500. The common markup for overhead and profit is 15 percent, or $101,325, bringing the total cost to $776,825. Given the uncertainty in the design of this project and the remote location, however, your company believes that an $850,000 fixed-price contract would ensure a net income of 15 percent.

Table B—Sample Estimates of Profit and Cost		
Type of Contract	**Contractor's Net Income**	**Owner's Cost**
Fixed-price– Owner pays $800,000.	$800,000 fixed-price – $675,500 cost ——————— $134,500 net income	$800,000 owner's cost
Cost-plus-fixed-fee– Contractor receives cost plus $85,000 fee.	$675,500 cost + $ 85,000 fee ——————— $760,500 – $675,500 cost ——————— $ 85,000 net income	$675,500 cost + $ 85,500 fee ——————— $760,500 owner's cost
Cost-plus-percentage-of-cost– Contractor receives cost plus 20 percent of cost.	$675,500 cost × .20 20 percent ——————— $135,100 net income	$675,500 cost + $135,100 20 percent ——————— $810,600 owner's cost
Incentive-Share overrun– Contractor receives cost plus fee minus one-half of cost overrun.	$877,025 actual cost – $776,825 estimated cost ——————— $100,200 overrun $ 85,000 fee – $ 50,100 one-half of overrun ——————— $ 34,900 income	$877,025 actual cost – $ 50,100 one-half of overrun ——————— $826,925 owner's cost
Incentive-Share savings– Contractor receives cost plus fee plus one-half of savings.	$766,825 estimated cost – $672,425 actual cost ——————— $ 94,400 savings – $ 47,200 one-half of savings + $ 85,000 fee ——————— $132,200 net income	$672,425 cost + $ 47,200 one-half of savings ——————— $719,625 owner's cost

Name _____

Table B illustrates how costs and profits vary based on the type of contract used. Substitute other values for estimated and actual cost and fees, to see how these changes affect net income and price. The construction company's overhead must be deducted from net income to calculate profit. Your group has 25 minutes to prepare your presentation for the client.

- Spend no more than 10 minutes deciding which type of contract you want to propose to the client.
- Spend no more than 5 minutes deciding the amount of the contract.
- Use the remaining time to prepare your presentation. Questions that may help you prepare your presentation include:
  - Will the type of contract you selected save the client money? How?
  - Does your contract reduce risk for the buyer? How?
  - Are you certain you will make a profit? Signing a contract that results in a loss could put you out of business.
  - What is the estimated profit?

Once all reports are given, the client will award a contract. Then answer the following questions:

1. Was your team awarded the contract?

   _____

2. If yes, why do you think your proposal was chosen?

   _____

   _____

3. If another team was awarded the contract, what did their proposal contain that was preferred by the client?

   _____

   _____

Chapter 5 Review
Construction Tools and Equipment

Name _____ Date _____ Class _____

Study Chapter 5 of the text, then answer the following questions.

1. List five tools that a plumber would bring to the job each day.

   _____
   _____
   _____
   _____
   _____

2. Name three tools that a carpentry contractor is likely to furnish for workers.

   _____
   _____
   _____

_____ 3. True or False? The best tool for a job is the least expensive tool.

_____ 4. Using a tool for a specified time period at a specified cost is known as ____.
   A. buying
   B. renting
   C. subcontracting
   D. All of the above.

_____ 5. A contract for a specific portion of the work necessary for completion of a project is called ____.

6. Name four types of trucks that are typically seen on a construction site.

   _____
   _____
   _____
   _____

Copyright by The Goodheart-Willcox Co., Inc.

7. Which of the following statements regarding shop areas is *not* true?
   A. Shop size will be dictated by the types of tools, materials, and equipment used in it.
   B. The shop is used mainly to do repair work on tools and equipment.
   C. Shop and storage areas are never combined into one large area.
   D. Some off-site work can be done in shop areas.

8. A secure, dry storage area is needed _____.
   A. for tools
   B. for equipment
   C. to deter theft
   D. All of the above.

9. True or False? Tools, materials, and supplies needed on a daily basis are stored in trucks or trailers.

10. _____ are materials used in the work process that do *not* become part of the finished product.

Name _____ Date _____ Class _____

Activity 5-1
Identifying Construction Equipment

Many types of tools and equipment are used on construction sites. Look through your textbook or on the Internet to find the name of the tool or machine shown in each box. Conduct additional research to learn how each item is used.

1.

Name and uses
_____
_____
_____
_____
_____

2.

Name and uses
_____
_____
_____
_____
_____

3.

Name and uses
_____
_____
_____
_____
_____

Chapter 5   Construction Tools and Equipment   45

Copyright by The Goodheart-Willcox Co., Inc.

46 Construction and Building Technology Tech Lab Workbook

4.

Name and uses

_____
_____
_____
_____
_____

5.

Name and uses

_____
_____
_____
_____
_____

6.

Name and uses

_____
_____
_____
_____
_____

7.

Name and uses

_____
_____
_____
_____
_____

Copyright by The Goodheart-Willcox Co., Inc.

Name _____

8.

Name and uses
_____
_____
_____
_____
_____

9.

Name and uses
_____
_____
_____
_____
_____

10.

Name and uses
_____
_____
_____
_____
_____

Name _____ Date _____ Class _____

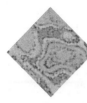

Activity 5-2
Tool Ownership

Imagine that you have accepted a job as a helper for an experienced tradesperson in one of the following job areas. You will work on single-family residential construction projects.

Step 1

Begin by circling the type of work you want to do:

Carpentry–framing

Roofing–shingles

Plumbing–rough-in

Electrical–rough-in

HVAC–rough-in

Copyright by The Goodheart-Willcox Co., Inc.

Step 2 _____

Complete the left column of Table A. Enter the names of the tools that you think will be needed in each of the categories. Use a pencil in case you need to change your answers based on the class discussion to follow.

Table A—Tool Ownership		
Tools Needed for Task	**Tool Size and Characteristics**	**Purchased Tool Size and Cost**
Measurement and Layout:		
Cutting and Drilling:		
Leveling and Plumbing:		
Assembly:		
Other:		

Step 3 _____

Participate in a class discussion regarding the tools used by different tradespeople. Review and modify your list based on this discussion. Complete the second column by identifying the desired size and characteristics of each tool.

Step 4 _____

Find prices for two or three tools. Visit a local retail outlet or look up the information online. Include the size and cost of the tools selected.

Chapter 6 Review
Concrete

Name _____ Date _____ Class _____

Study Chapter 6 of the text, then answer the following questions.

_____ 1. Concrete _____.
   A. can be strengthened with steel reinforcing bars
   B. is used for footings, foundations, and walls
   C. is made of three ingredients
   D. All of the above.

Match each type of Portland cement with the correct description.

_____ 2. Type I
_____ 3. Type II
_____ 4. Type III
_____ 5. Type IV
_____ 6. Type V

   A. Low heat, strength develops more slowly.
   B. Moderate sulfate resistance, used when surrounding soil contains elevated levels of sulfate.
   C. High, early strength, used when curing must be done quickly.
   D. Adds color.
   E. General purpose, used when special properties are not required.
   F. Sulfate resistant, used in areas where soil contains extreme levels of sulfate.

_____ 7. True or False? Aggregate consists of sand and Portland cement.

8. Why are admixtures added to concrete?
_____
_____

_____ 9. True or False? Concrete is much stronger in tension than in compression.

_____ 10. The most important _____ of concrete include strength, watertightness, durability, and workability.
   A. qualities
   B. quantities
   C. properties
   D. admixtures

11. Review the following steps involved in working with concrete, then place them in the order in which they are completed.

   _____ Finishing
   _____ Curing
   _____ Preparing subgrade and forms
   _____ Obtaining concrete
   _____ Placing
   _____ Estimating volume

Identify the lettered parts of the concrete wall form in the following drawing.

   _____ 12. Form ties
   _____ 13. Brace
   _____ 14. Rebar
   _____ 15. 2 × 4 or metal frame
   _____ 16. Stake
   _____ 17. Plywood
   _____ 18. Clamps
   _____ 19. 2 × 4 whalers

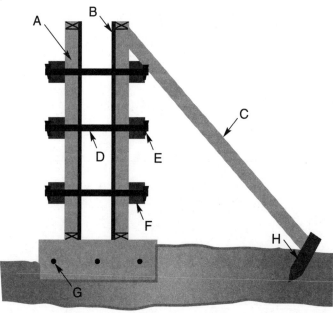

Name _____

20. Calculate the volume of concrete needed for a patio that will be 4″ × 15′ 6″ × 21′ 9″. Show the general formula you used and your calculations.

_____ 21. During _____, fresh concrete is compacted to fill all voids and air pockets inside a concrete form.

Match each of the following concrete finishing terms with the correct definition.

_____ 22. Jointing
_____ 23. Darbying
_____ 24. Floating
_____ 25. Screeding
_____ 26. Edging
_____ 27. Troweling

A. Creating a radius along the edge of a concrete slab to prevent chipping and to improve appearance.

B. A process that removes excess concrete and brings the top surface to the proper grade.

C. Making shallow grooves in concrete to control cracking.

D. A process that produces a dense, smooth, hard surface.

E. A chemical process that causes concrete to become a solid.

F. A process that removes imperfections and prepares the surface for the final finish.

G. A process that levels ridges or fills voids left after screeding.

_____ 28. In order for concrete to harden properly, _____ is required.

_____ 29. True or False? A slump test measures the consistency of batches of concrete.

_____ 30. The ability of concrete to support a load is measured by a(n) _____ test.

Name _____ Date _____ Class _____

Activity 6-1
Concrete Tools

Using the Internet, research the concrete tools listed in the table. Information about the bull float has been completed for you to use as a guide.

\	Concrete Tools					
Name	Description	Size	Power	Special Features	Material	Cost
Bull float	Flat aluminum channel with bracket for long handle	8″ × 36″ 8″ × 48″	Hand	Some have rounded ends	Magnesium Aluminum	$84.35 $79.50
Mixing hoe						
Float						
Concrete saw						
Finishing machine						
Form scraper						
Concrete vibrator						
Edger						
Groover						
Trowel						
Darby						
Tamper						

Copyright by The Goodheart-Willcox Co., Inc.

Chapter 7 Review
Metals

Name _____ Date _____ Class _____

Study Chapter 7 of the text, then answer the following questions.

Match the forming process with the correct description.

_____ 1. Casting
_____ 2. Drawing
_____ 3. Extruding
_____ 4. Rolling
_____ 5. Forging

A. Heating to a high temperature and passing between sets of rollers.

B. Forcing steel through dies of the appropriate shape.

C. Spinning liquid steel into threads.

D. Pouring molten metal into molds.

E. Pressing or stamping preheated steel between dies of the desired shape.

F. Pulling steel through a series of dies to produce a desired cross-sectional shape.

_____ 6. True or False? Structural steel is the same thing as bar stock.

Identify each type of cross-sectional steel shape shown.

_____ 7.
_____ 8.
_____ 9.
_____ 10.
_____ 11.
_____ 12.

_____ 13. Large, horizontal structural members supported by columns are called ____.
   A. beams
   B. girders
   C. columns
   D. joists

14. Name two methods for joining structural steel pieces together.
_____
_____

15. The basic steps for erecting a structural steel are listed here. Indicate the order in which they are done using the letters A-E.

_____ The frame is plumbed, leveled, and temporarily braced.

_____ Beams are attached to girders.

_____ Permanent bracing is installed.

_____ Columns are lifted into placed and bolted to the foundation.

_____ Girders are attached to columns

_____ 16. Bent pieces of wire or plastic that hold reinforcing material above the ground or a form are known as ____.

_____ 17. Metal material used to reinforce concrete floors and paving is known as ____ wire mesh.

18. Explain the major difference between steel plate and sheet steel.
_____
_____

19. List three advantages metal framing has over wood framing.
_____
_____
_____
_____
_____

20. Name two types of framing connectors.
_____

Name _____

21. Name two types of nails and how each type is typically used.

   _____
   _____
   _____
   _____

_____ 22. Fasteners that are inserted through holes drilled into the material being fastened are called _____.
   A. nails
   B. bolts
   C. screws
   D. strapping

23. List five characteristics that are used to classify screws.

   _____
   _____
   _____
   _____
   _____

_____ 24. True or False? In face nailing, nails are driven at a right angle to the surface of the board being nailed.

25. Name four products used in construction projects that are made from aluminum or copper.

   _____
   _____
   _____
   _____
   _____

Name _____ Date _____ Class _____

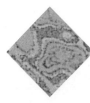
Activity 7-1
Metal Materials

In this activity, your instructor will show you a variety of metal materials and discuss how they are used, their available sizes, and cost. Complete the table as each item is discussed. There is room at the end of the table for additional materials your instructor may show.

Metal Construction Materials				
Material	Uses	Available in Lab?	Sizes	Cost
Rebar				
Wire mesh				
Black iron pipe				
Galvanized pipe				
90° elbow				
Tee				
Coupling				
Common nail				
Box nail				
Finish nail				
Flat head wood screw				
Round head wood screw				
Pan head screw				
Carriage bolt				
Machine bolt				
Aluminum gutter				
Aluminum downspout				
Aluminum siding				

Name _____ Date _____ Class _____

Activity 7-2
Simple Life-Cycle Cost Analysis

A life-cycle assessment determines the environmental and monetary costs associated with the entire life cycle of a process or a product. A life-cycle cost analysis, however, estimates the total dollar cost of owning, operating, maintaining, and disposing of a product for a specified time. In this activity, you will determine the costs of buying, owning, and reselling a pick-up truck for a specified time.

Imagine that you want to purchase a 2 1/2-year-old pick-up truck. The truck you have found is in excellent condition, with 25,000 miles on the odometer. The purchase price is $25,000. You plan to own the truck for five years, drive it an average of 12,000 miles per year, and average 18 miles per gallon of fuel. After five years, the trade-in value for the truck is expected to be $10,000, provided the truck is kept in good condition.

In the table below, estimate the total cost of owning and operating the truck for five years, using the values given.

Life-Cycle Cost Analysis		
Material	**Rate per Unit**	**Total Cost**
Purchase Price	$25,000	$25,000
Operating costs		
A. License and insurance	$1150 per year	A. =
B. Fuel	$4.50 per gallon	B. =
C. Oil and filter change every 4000 miles	$40	C. =
D. Tire replacement, four tires, every 30,000 miles	$100 per tire	D. =
E. Battery replacement, twice	$130	E. =
F. Brake pad/shoes replacement, twice	$250 per replacement	F. =
G. Miscellaneous—bulbs, fluids, wiper blades, etc.	$20 per year	G. =
	Add total costs for subtotal =	
	Minus trade-in value	− $10,000
	Total cost	
	Total cost per mile	

Copyright by The Goodheart-Willcox Co., Inc.

Chapter 8 Review
Wood and Wood Products

Name _____ Date _____ Class _____

Study Chapter 8 of the text, then answer the following questions.

_____ 1. Deciduous trees are the source of _____ lumber.

_____ 2. True or False? Hardwoods are more dense than softwoods.

3. Using the letters A–I, place the steps in the lumber production process in the order in which they are completed.

_____ Remove limbs

_____ Transport logs to sawmill

_____ Saw to desired thickness

_____ Cut into logs

_____ Season lumber

_____ Debark

_____ Grade lumber

_____ Cut down trees

_____ Wash

_____ 4. Softwood lumber that is less than 2" thick and graded based on appearance is classified as _____.
   A. dimension lumber
   B. good-one-side
   C. light framing lumber
   D. boards

_____ 5. Which of the following statements about dimension lumber is *not* true?
   A. It is commonly used as a structural component.
   B. It is more than 2" but less than 5" in nominal thickness.
   C. It is graded based on the appearance of the board face.
   D. Typical grades include light-framing, stud, and structural light-framing.

_____ 6. Graders estimate the strength of a piece of lumber based on the number, size, type, and location of _____ found on the lumber.

_____ 7. The initials MSR stand for machine _____ rated.

_____ 8. True or False? Hardwood lumber is sold only in random widths and lengths.

_____ 9. The actual width dimension of dimension lumber is approximately _____ inch less than the nominal dimensions.
   A. 1/2
   B. 1/8
   C. 3/8
   D. 1/4

_____ 10. The number of pounds of preservative in one cubic foot of wood is known as the _____ level.
   A. saturation
   B. absorption
   C. retention
   D. treatment

11. List five safety rules to adhere to when working with lumber.

_____
_____
_____
_____
_____

_____ 12. Wood panel products are _____ materials, made by binding wood strands, particles, fibers, or veneers with adhesives.

_____ 13. An exposure durability rating indicates a plywood panel's resistance to _____.
   A. weather
   B. hammer blows
   C. paint
   D. sunlight

Name _____

Match the following plywood terms with the correct description.

_____ 14. Good-one-side

_____ 15. High-density overlay

_____ 16. Good-two-side

_____ 17. Medium-density overlay

A. Has an easy-to-paint surface.

B. High-quality veneer on one face and lower quality veneer on the back.

C. Has high-quality veneer on the front and back faces.

D. Exterior-grade plywood.

_____ 18. Which of the following statements about hardboard is true?
  A. Made from long, narrow, strand-like particles.
  B. Has two smooth, grain-free surfaces.
  C. Has large, randomly oriented particles.
  D. Is dense and abrasion resistant.

19. Explain how medium-density fiberboard is manufactured.

_____

_____

_____

20. Explain the main difference between oriented strand board and waferboard.

_____

_____

_____

_____ 21. True or False? Engineered lumber and beams are not as consistently strong as sawn wood products.

Match the wood product with the correct description.

_____ 22. Laminated-veneer lumber
_____ 23. Glue-laminated beams
_____ 24. Wood I-joists
_____ 25. Open-web joists
_____ 26. Board and batten
_____ 27. Wood shakes
_____ 28. Wood molding
_____ 29. Wood shingles

A. Long strips of wood used for baseboards and window and door trim

B. Made with plywood or OSB webs and LVL top and bottom cords.

C. Cords are made of high-quality 2 × 4 lumber, webs are made of wood or steel.

D. A method of installing boards on a building's exterior to form vertical siding.

E. Made using 1/8" × 1/2" × 8' veneer strands.

F. Taper sawn wood product used for roofing and siding.

G. Made using 1/8" thick by 8' long strips of veneer.

H. Made by bonding together several 2" thick pieces of lumber.

I. Hand split wood product used for roofing and siding.

Name _____ Date _____ Class _____

Activity 8-1
Wood Materials

In this activity, your instructor will show you a variety of wood materials and discuss how they are used, their available sizes, and cost. Complete the table as each item is discussed. There is room at the end of the table for additional materials your instructor may show.

Wood Construction Materials				
Material	**Uses**	**Available in Lab?**	**Sizes**	**Cost**
Boards				
Dimension lumber				
Plywood				
Medium-density overlay (MDO)				
Hardboard				
Medium-density fiberboard (MDF)				
Oriented strand board (OSB)				
Waferboard				
Laminated-veneer lumber				
Glue-laminated beams				
Wood I-Joist				
Open web joist				
Parallel-strand lumber				
Wood shingles and shakes				
Horizontal wood siding				
Vertical wood siding				
Baseboard				
Cove molding				
Crown molding				

Copyright by The Goodheart-Willcox Co., Inc.

Chapter 9 Review
Masonry, Glass, and Plastics

Name _____ Date _____ Class _____

Study Chapter 9 of the text, then answer the following questions.

_____ 1. The process of building with stone, brick, and similar materials is known as _____.

Match the type of stone listed with the correct description.

_____ 2. Granite

_____ 3. Sandstone

_____ 4. Limestone

_____ 5. Marble

_____ 6. Slate

_____ 7. Manufactured stone

A. Dense, fine-grained stone used for floor tile and roofing.

B. Sedimentary stone composed of nearly pure silica.

C. Hard and durable, can be ground to a highly polished surface.

D. Made from lightweight concrete.

E. A type of limestone that can take a high polish.

F. Stone created from organic materials, made primarily of calcite.

_____ 8. A structural product made from clay or shale and molded, then fired is known as _____.
   A. mortar
   B. stone veneer
   C. brick
   D. granite

9. What is a kiln and what is it used for?

   _____

10. Name seven types of bricks.

    _____
    _____
    _____
    _____

_____ 11. True or False? Facing brick has a large hollow core.

12. What are the two size categories for brick?
_____
_____

_____ 13. The dimensions of _____ brick do not include a mortar joint as part of the size.

_____ 14. The parts of concrete block can include _____.
　　A. cores
　　B. plain ends
　　C. ears
　　D. All of the above.

15. In addition to water, what are the four ingredients used to make mortar?
_____
_____
_____
_____

16. Identify the steps involved in laying masonry units in the order they are normally completed.
_____
_____
_____
_____

_____ 17. Safety glass that is made by heating glass to near its melting point and then cooling it rapidly is known as _____ glass.

_____ 18. When _____ glass is broken, the pieces of glass are held together by a plastic film.
　　A. window
　　B. coated
　　C. insulated
　　D. laminated

_____ 19. True or False? The invention of wired glass made it possible to install panes of glass in fire doors.

_____ 20. The coating on the inner layer of _____ glass makes the window more energy efficient.

Name _____

Match the type of plastic with the correct description.

_____ 21. Polycarbonate plastic

_____ 22. Polyvinyl chloride

_____ 23. Chlorinated polyvinyl chloride

_____ 24. Polyethylene

_____ 25. Styrene rubber

_____ 26. Vinyl

A. Used to make corrugated pipe and fittings.

B. Suitable for hot water supply.

C. Used primarily as a replacement for glass.

D. Used to make siding, soffit, fascia, gutters, and floor coverings.

E. Used for cold water supply and drain, waste, and vent piping.

F. Used to pipe natural gas.

G. Used for both hot and cold water supplies.

Name _____ Date _____ Class _____

Activity 9-1
Masonry, Glass, and Plastic Identification

In this activity, your instructor will show you a variety of masonry, glass, and plastic materials and discuss how they are used, their available sizes, and cost. Complete the table as each item is discussed. There is room at the end of the table for additional materials your instructor may show.

Masonry, Glass and Plastic Materials				
Material	**Uses**	**Available in Lab?**	**Sizes**	**Cost**
Granite				
Sandstone				
Limestone				
Marble				
Slate				
Common brick				
Building brick				
Face brick				
Plate glass				
Stretcher block 8 × 8 × 16				
Stretcher block 4 × 8 × 16				
Half-block 8 × 8 × 8				
Concrete brick				
Single-strength glass				
Double-strength glass				
Tempered glass				
Laminated glass				
Wired glass				
Insulated glass				
Reflective glass				
Glass block				
Acrylic plastic sheet				
Polycarbonate plastic sheet				

(Continued)

Copyright by The Goodheart-Willcox Co., Inc.

Masonry, Glass and Plastic Materials (*Continued*)				
Material	Uses	Available in Lab?	Sizes	Cost
Polyvinyl chloride (PVC)				
Chlorinated polyvinyl chloride (CPVC)				
Cross-linked polyethylene (PEX)				
Polyethylene (PE)				
Styrene rubber (SR)				
Vinyl siding				
Vinyl soffit and fascia				
Vinyl gutters and downspouts				
Vinyl floor coverings				

Chapter 10 Review
Architectural Design

Name _____ Date _____ Class _____

Study Chapter 10 of the text, then answer the following questions.

_____ 1. True or False? Designing is a process for developing and evaluating solutions to a problem.

2. Indicate whether an architect or engineer typically completes each of the following design activities. Use an *A* to indicate architect, *E* to indicate engineer.

   _____ Pay close attention to the size, shape, and relationship of spaces

   _____ Design highways, bridges, and dams.

   _____ Solve design problems related to building structures.

   _____ Consider appearance an important part of the design.

   _____ Design utility systems.

   _____ Evaluate the relationship of a building to a site.

3. Architects keep five design factors in mind as they create and implement a design. Name these factors.

   _____

   _____

   _____

   _____ 4. Appearance as a design consideration includes _____.
   - A. only the interior of a building
   - B. considers the relationship of the building to its surroundings
   - C. only the exterior of the building
   - D. None of the above.

5. Identify the missing steps of the design process in the following figure.

   A. _____
   B. _____
   C. _____
   D. _____

   ```
   A
   ↓
   Generate preliminary ideas
   ↓
   B
   ↓
   Analyze ideas
   ↓
   C
   ↓
   D
   ↓
   Implement design
   ```

6. What are preliminary ideas?

   _____
   _____
   _____

   _____ 7. True or False? Elements from two or more preliminary ideas should never be combined to create a workable solution to a design problem.

Match the following types of analysis with the correction description.

   _____ 8. Functional analysis
   _____ 9. Cost analysis
   _____ 10. Site analysis
   _____ 11. Structural analysis

   A. Evaluation of the strength required of a proposed structure.
   B. Evaluation of expenses involved in various stages of a construction project.
   C. Evaluation of the usefulness and practicality of a proposed structure.
   D. Evaluation of how structures will fit on a particular site.

Name _____

_____ 12. Which of the following actions is *not* done during the design selection phase of the design process?
   A. Architects meet with the client.
   B. Drawings, models, cost estimates and other supporting materials are prepared.
   C. Construction documents are prepared.
   D. The client accepts a design, asks for modifications, or abandons the project.

_____ 13. All working drawings and written specifications used in the construction of a structure are known as _____ documents.

_____ 14. During design implementation, _____.
   A. the client approves of plans and specifications
   B. contractors are hired
   C. ends with project completion
   D. All of the above.

Name _____ Date _____ Class _____

Activity 10-1
Reading Residential Drawings

Architects, engineers, contractors, and skilled workers use working drawings to build a project. Your instructor will provide you with a set of residential working drawings. Working with a group of classmates, answer the following questions about the drawings. Group participation and cooperation are essential to successfully complete this activity.

1. Find the title block on the first page of the drawings. It is typically located along the bottom edge of each page of the drawings.

    A. What is the title or name of the drawings?

    _____

    B. Who was the home designed for?

    _____

    C. Who prepared the drawings?

    _____

2. Find the plot plan.

    A. What is the scale of this drawing?

    _____

    B. If the property borders a street or road, what is its name?

    _____

    C. What is the maximum width of the lot? Depth?

    _____

    D. What is the minimum distance from the street to the front of the building?

    _____

    E. What is the maximum width of the building? Depth?

    _____

Copyright by The Goodheart-Willcox Co., Inc.

3. Find the foundation plan. Locate the walls around the perimeter of the house. Look for section lines along the perimeter walls like those shown below.

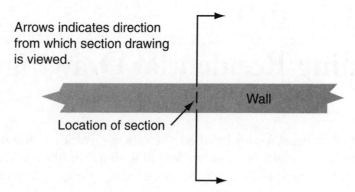

A. What letters or numbers identify the sections?

B. Locate the pages that contain section drawings. Find the section drawing that corresponds to the labels found in Question 3A. What is the name of this drawing?

C. What is the thickness of the foundation wall at this point?

D. What materials will be used to build the walls?

E. What is the footing width? Height?

4. Return to the foundation plan found in Question 3.

A. Are footings shown inside the perimeter walls? If yes, are they located in a grid pattern? If no, does a footing run the length or width of the foundation, approximately midway between two exterior walls?

B. What is the footing width? Length?

C. What purpose are these footings likely to serve?

5. Find the first floor plan.

A. What is the scale of this drawing? What fraction of actual size is this drawing?

B. What is the largest room on this floor?

Name _____

    C. How many rooms are located on this floor?
    _____

    D. How many restrooms are located on this floor?
    _____

    E. How many staircases lead to other floors?
    _____

    F. How many exterior doors are located on this floor?
    _____

    G. How many windows are located on this floor?
    _____

    H. What is the rough opening size of the widest window? (Hint: Look for a table labeled *Rough Openings*.)
    _____

6. Find the front elevation plan.

    A. How many floors are aboveground?
    _____

    B. What materials are visible on the front of the building?
    _____

    C. How many entrances are on this side of the building?
    _____

    D. How many windows are in this elevation?
    _____

7. If the house has a second floor, find the second floor plan and answer the following questions. If not, proceed to Question 8.

    A. What is the largest room on the second floor?
    _____

    B. How many rooms are located on this floor?
    _____

    C. How many restrooms are located on this floor?
    _____

    D. How many staircases lead to other floors?
    _____

E. How many exterior doors are located on this floor?

F. How many windows are located on this floor?

G. What is the rough opening size of the narrowest window?

8. Find the right-side elevation plan.

    A. What materials are visible on this side of the building?

    B. How many entrances are on this side of the building?

    C. How many windows are in this elevation?

9. Is there a separate plumbing drawing in this set of plans? If so, use it to answer the following questions. If not, refer to the floor plans.

    A. Locate the kitchen. What plumbing fixtures are shown in this room?

    B. Record the overall dimensions of each of the restrooms on this floor. List the fixtures found in each in the table below.

Dimensions	Fixtures

    C. Are plumbing fixtures included in any other rooms on this floor? If yes, list the names of the rooms and the fixtures included. (Hint: Look for a laundry room or wet bar.)

Room	Fixtures

Chapter 10   Architectural Design   85

Name _____

10. If the house has a second story, answer the following questions.

   A. Record the overall dimensions of each of the restrooms on this floor. List the fixtures found in each in the table below.

Dimensions	Fixtures

   B. Are plumbing fixtures included in any other rooms on this floor? If yes, list the names of the rooms and the fixtures included.

Room	Fixtures

Use the electrical plans, floor plans (if there are no electrical plans), and the electrical symbols shown here to answer Questions 11 and 12.

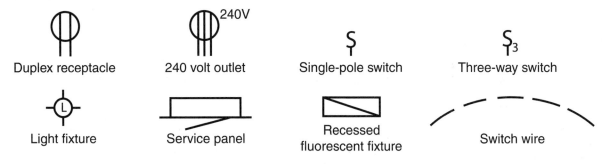

11. Locate the master bedroom.

   A. How many duplex receptacles are in this room?

   _____

   B. How many single-pole switches are in the master bedroom?

   _____

   C. What lights or duplex receptacles do each of these switches control?

   _____

   D. How many three-way switches are in this room?

   _____

   E. What lights or duplex receptacles or lights do these switches control?

   _____

Copyright by The Goodheart-Willcox Co., Inc.

12. Locate the kitchen.

    A. How many duplex receptacles are in this room?

    B. Is a 220-volt receptacle provided for a range?

    C. How many single-pole switches are in the kitchen?

    D. What lights or duplex receptacles do each of these switches control?

    E. How many three-way switches are in this room?

    F. What lights or duplex receptacles do these switches control?

13. Find the service entrance, garage, utility room, and basement.

    A. Where is the service entrance located?

    B. Is the type of service entrance indicated? If so, what type is required?

14. Find the section drawings.

    A. Select one of the section drawings other than the one you identified earlier. What is the title of this section?

    B. What is the scale of this drawing? What fraction of actual size is this drawing?

    C. What part of the home is shown in this section drawing?

    D. Where in the plan is this section located? Provide page number and the title of section drawing.

Name _____

   E. Does this drawing help you understand how some parts of the house will be assembled? Why or why not?

   _____
   _____
   _____

15. Find the detail drawing for selected parts of the interior. These drawings often indicate the appearance of cabinetry, fireplace mantels, built-in bookcases, and other interior finish work.

   A. What detail drawings did you find?

   _____

   B. What is the scale of one of these drawings?

   _____

   C. Are notes and dimensions included with this drawing?

   _____

Name _____ Date _____ Class _____

Activity 10-2
Green Certification Programs

This activity introduces two widely known green certification programs in construction technology: Leadership in Energy and Environmental Design (LEED) and the National Green Building Standard (NGBS).

Working in two teams, one team will report on LEED certification and the other team, NGBS certification. Each member of the team should prepare a two-minute presentation about an important characteristic of the assigned green certification program. Preparation of individual presentations will be done outside of class, but your instructor will schedule class time for team meetings.

During the first meeting, select a team leader. The team leader will coordinate the work of the team members. Additional responsibilities of the team leader include:

- making certain team members understand their responsibilities.
- providing the instructor with a list of the topics to be presented by each member of the team.
- clarifying boundaries between topics to avoid excessive overlap in presentations.
- providing leadership in team meetings.

As a team, assign topics to each team member. Remember to include an introduction and summary. Your team might wish to include some of the following topics in the presentations:

- the purpose and history of the program.
- the categories used in the rating system.
- requirements for each level of certification.
- organizations involved in developing and operating the program and the function of each organization.
- steps involved in process of becoming certified.
- educational programs provided and credentials awarded.
- examples of homes certified by the program.
- examples of builders constructing certified homes.
- examples of design companies that design certified homes.

Use the time between the first and second meeting to research your topic. The Internet is a good source for information. Begin by searching for your assigned certification program. Search within those results for your assigned topic.

During the second meeting, decide on the length of each presentation. Decide if any teaching aids, such as reports or outlines, will be used during presentations. Before the third meeting, team members should time their presentations and, if needed, modify them so they last no longer than two minutes. At the third meeting, develop a schedule to move the presentation along, with little delay between presenters. For example, seating presenters in order of their presentation saves time, as does having team members help each other with teaching aids.

Important dates:

First team meeting _____

Second team meeting_____

Third team meeting _____

Presentation date _____

Activity 10-3
Designing a Storage Building

Part 1

In the next four activities, you will follow the steps in the design process to design a storage building. You will work in groups to complete each part of the activity. Part 1 focuses on preparing a floor plan. In Part 2 you will work on the appearance of the storage building. In Part 3, each team will analyze their own designs and prepare a presentation for the class. Finally, in Part 4 the entire class will review all of the designs and select one to be constructed. Your instructor will tell you how much class time will be available for each major element of the activity. The tools and materials needed for Part 1 of this activity include:

- Scissors
- 12″ ruler
- Pencil
- Eraser

Identify the Problem

Identifying the problem is the first step in the architectural design process. In this instance, the problem is to design a storage building for yard and garden tools and equipment. Requirements include adequate space to store all items and an exterior appearance that fits in with local residential architecture.

Space requirements include the following:

- Floor space for a lawn mower, garden cart, fertilizer spreader, two large plastic containers for soil supplements and fertilizer, and a 6′ stepladder.
- Wall space to hang rakes, shovels, brooms, a string trimmer, and two 50′ garden hoses.
- Shelves to hold pots, garden chemicals, a tank sprayer, and other small items.
- Space to access most items without needing to move items stored on the floor.

Requirements of the building's appearance include:

- Compatibility with residential architecture in the area.
- A durable, easily maintained exterior finish.
- Roof, wall, and trim paint colors that are appropriate for a garden setting.

Generate Preliminary Ideas

During this step, you will develop various floor plans. Find the templates provided at the end of this activity. Cut them apart and use the pieces to develop your floor plans.

1. Select the templates for items that will be stored on the floor and for the access area. Using the graph paper labeled "Floor Plan" at the end of this activity, try various arrangements of these templates until you have an efficient layout.

2. Locate the tool panel template. The tool panel will be 6" × 72" and hung on the wall at a height of at least 6′. This provides floor space for tools that will hang on the surface of the panel and extend to the floor. The tool panel can be cut into two pieces so it can be hung in a corner or on two walls.

3. The access template represents a 30" wide area for entering and moving around in the building. There is enough open space to get within 3′ of the back wall without moving anything. Cut the template to length once the final layout is determined.

4. Make any desired adjustments in the layout. Center the layout within the border of the graph paper.

5. Using a ruler as a guide, draw light, dashed lines around each of the templates.

6. Pressing lightly on your pencil, draw a rectangle around all of the templates. This rectangle represents the inside of the exterior walls.

7. Lightly draw a second solid-lined rectangle outside and 4" away from the rectangle drawn in Step 6. Each square represents 3 inches. This rectangle represents the outside of the exterior walls.

8. Locate the door so that the access template is inside the door. You can make the door wider than 30" to allow equipment to be moved into and out of the building. Use one of the door symbols shown here:

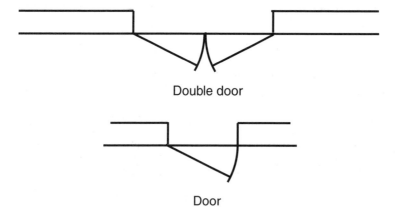

Double door

Door

9. Erase any unwanted lines and darken the solid lines for the interior and exterior of the walls and door. Darken the dashed lines that were drawn in Step 5.

10. Locate the shelving templates. Two shelving templates are marked optional. You do not have to include them in the plan. If you have a space where they can fit, however, you may include all or part of them. You may also cut the shelving to lengths that fit your layout and join pieces end to end to make long shelves. The only items stored on the floor that are over 4′ tall are the ladder and the large tool panel. Therefore, shelves can be installed one above the other over shorter items stored on the floor.

Name _____

11. Once you have placed the shelving in the plan, use the ruler as a guide to draw solid lines around the shelving templates. In locations where you plan to install one shelf above another, label the shelves as either *Shelf—12″ × (length)* or *Two shelves—12″ × (length)*.

12. Add labels inside each item drawn in Step 5.

13. Add extension and dimension lines, and overall dimensions lines for the entire structure. Make the lines as shown in the following drawing:

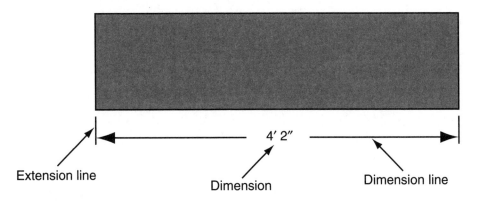

14. Add your name and today's date in the title block.

15. What advantages does this floor plan offer? Consider the following questions as you consider the advantages:
    - Does this plan provide space to store all the items specified?
    - Is the space used efficiently?
    - Is the amount of access space minimized?

    Explain the advantages of this plan on the back of the drawing.

Refine Ideas

Have each member of your team present their storage building floor plan design. Following the presentations, decide the overall dimensions of the floor frame. Your group may select one of the plans or combine components of several plans. The group may also decide to increase the overall size floor frame to take maximum advantage of standard 4′ × 8′ sheets of plywood used for the flooring. The dimensions chosen will be used to create elevation drawings in Activity 10-4.

Name _____

Templates

Lawn mower
30″ × 42″

Garden cart
24″ × 36″

Fertilizer spreader
18″ × 30″

Access 30″ wide
length veries

Tool panel 6″ × 72″

Optional tool panel

Stepladder

Plastic container 24″ Dia.

Plastic container 24″ Dia.

Shelf
12″ × 12′ 0″

Shelf
12″ × 12′ 0″

Optional Shelf
12″ × 12′ 0″

Name _____

Drawn by:	Date:	**Floor Plan**	**Storage Building**

Name _____

Drawn by:	Date:		**Storage Building**

Name _____ Date _____ Class _____

Activity 10-4
Designing a Storage Building

Part 2

In this activity, you will create designs for the exterior of the storage building. Begin by reviewing the problem statement in Activity 10-3. You will need the following tools and supplies:

- 12" ruler
- Pencil
- Eraser

Generate Preliminary Ideas

Recall that the width of each elevation was determined in the Part 1 of the activity. The team will now develop ideas for the exterior appearance of the storage building.

1. Prepare a front elevation (elevation with the door) and side elevation of the storage building. The roof design and exterior appearance must be compatible with residential architecture in the area. Exterior elevation requirements include:
   - Minimum ceiling height of 7' 0".
   - Minimum roof slope of 4 in 12.
   - Floor frame thickness of 6".
   - Floor frame support will be specified by your instructor.

2. Sketch your ideas on a separate sheet of paper before drawing them on graph paper.

3. Prepare the front elevation, using the graph paper marked "Front Elevation" at the end of this activity. Each square represents 3 inches. Center the drawing within the borders of the paper. Draw light lines so they can be erased. Once all lightly drawn lines are correct, darken the lines.

4. Add overall dimensions for width and height.

5. Repeat Steps 2–4 to prepare the side elevation. Use the graph paper marked "Side Elevation" at the end of this activity.

6. Why do you believe your elevations should be selected for the storage building?

_____

_____

_____

Copyright by The Goodheart-Willcox Co., Inc.

Refine Elevation Ideas

7. Each team member will present their idea to the team. Following the presentations, discuss the designs and consider the following questions. How can individual designs be improved? Can several designs be combined to provide a better solution? What type of materials will be used for each component shown in the elevations?

8. Select two sets of elevations for the team's storage building design. These may be individual designs or combinations of several designs.

Name _____

Drawn by:	Date:	**Front Elevation**	**Storage Building**

Name _____

| Drawn by: | Date: | | **Storage Building** |

Name _____

Drawn by:	Date:	**Side Elevation**	**Storage Building**

Name _____

Drawn by:	Date:		**Storage Building**

Name _____ Date _____ Class _____

Activity 10-5
Designing a Storage Building

Part 3

In this activity, you will analyze the designs you have developed, prepare the initial draft of the specifications for the storage building, and prepare the group presentation of the design for your class.

Analyze Ideas

Based on the fourth step in the design process, you will analyze the floor plan and the two sets of elevations chosen by your group for any additional improvements. Once your team is satisfied with the designs, you will select one set of elevations to present to the class.

1. Working individually, review the space requirements outlined in the "Identify the Problem" section of Activity 10-3 Then answer the following questions regarding the floor plan.

   A. Does the plan provide space to store all of the items specified? If not, what changes are needed?

   _____
   _____
   _____

   B. Does the plan include floor space that is not needed? Can this space be eliminated? If so, how?

   _____
   _____
   _____

   C. Is the amount of floor space provided for access adequate? If not, what modifications do you recommend?

   _____
   _____
   _____

   D. Could the amount of space allocated for access be reduced? If so, how?

   _____
   _____
   _____

E. Are the overall dimensions such that 4′ × 8′ plywood sheets can be efficiently used for the flooring? If not, what overall dimensions would work?

_____
_____
_____

F. Respond to any other suggestions made by your instructor about the floor plan.

_____
_____
_____
_____
_____

2. Discuss the responses to the previous questions with your group. Briefly describe any changes the group will make to the floor plan.

_____
_____
_____
_____

3. Individually review the building appearance requirements identified in Activity 10-3. Answer the following questions regarding the elevation drawings:

A. What changes do you recommend to improve the building's appearance or to make it more compatible with homes in the area?

_____
_____
_____

B. What type of materials do you recommend for the exterior walls and trim?

_____

What type of shingles do you recommend for the roof?

_____

What colors do you recommend for the exterior walls and siding?

_____

4. Discuss the responses to the above questions with your group. Then decide which set of elevation drawings to present to the class.

Name _____

5. Briefly describe any changes the group will make to the elevation drawings selected in Step 4.

   _____
   _____
   _____
   _____

6. Choose three team members to modify the floor plan and elevations according to the changes chosen in Steps 2 and 4.

7. Prepare your class presentation of the design your team has developed. Decide what design features to emphasize, how to organize the presentation, and who will be responsible for each part of the presentation.

Name _____ Date _____ Class _____

Activity 10-6
Designing a Storage Building

Part 4

Step 1

In the fifth step of the design process, the class will choose one storage building design to build. Each design team will present their plans for consideration by the class. Use the following directions as a guide through the selection process.

1. Review the Ranking Form that follows Question 3. Keep these assessments in mind as you listen to each presentation.

2. Listen carefully to each team's presentation and participate in the discussion that follows.

3. Complete the form following each presentation. Summarize the key features, strengths, and weaknesses of each design.

Design team	
Key features	
Strengths	
Weaknesses	

Design team	
Key features	
Strengths	
Weaknesses	

Design team	
Key features	
Strengths	
Weaknesses	

Design team	
Key features	
Strengths	
Weaknesses	

Name _____

Design team	
Key features	
Strengths	
Weaknesses	

Design team	
Key features	
Strengths	
Weaknesses	

Design team	
Key features	
Strengths	
Weaknesses	

Copyright by The Goodheart-Willcox Co., Inc.

4. When all the presentations are completed, rank the designs. Number 1 indicates the best proposal. Record your rankings in the following table. Be prepared to discuss your rankings.

Team	Rank
A	
B	
C	
D	

Chapter 11 Review
Construction Engineering

Name _____ Date _____ Class _____

Study Chapter 11 of the text, then answer the following questions.

_____ 1. A process for developing and evaluating solutions to construction problems is called ____.
    A. architectural engineering
    B. contract engineering
    C. construction engineering
    D. construction management

2. Define the term *engineer*.

_____
_____
_____

3. Name five types of engineers.

_____
_____
_____

_____ 4. True or False? Engineers do not focus on the same factors that architects focus on when designing a project.

5. The work of a(n) ____ is more scientific and mathematical and less artistic than the work of a(n) ____.

_____

_____ 6. In nearly all types of construction engineering, ____ is a critical factor.

_____ 7. Engineers select ____ based on their strength, weather and fire resistance, cost, and appearance.

_____ 8. In addition to materials and budget, project costs are affected by _____.
   A. site choice
   B. structural design
   C. construction methods
   D. All of the above.

_____ 9. True or False? Engineers place less emphasis on design appearance than do architects.

Match the step in the engineering process with the correct description of what occurs during this step.

_____ 10. Identify the problem.
_____ 11. Generate preliminary ideas.
_____ 12. Refine ideas.
_____ 13. Analyze ideas.
_____ 14. Select the design.
_____ 15. Prepare construction documents.
_____ 16. Implement design.

A. Expand on some preliminary ideas.
B. Prepare detailed drawings and specifications.
C. Construction work begins.
D. Observe, investigate, and gather information.
E. Evaluate function, site, structure, and cost of each idea.
F. Consider any and all ideas.
G. Grant permission to engineers to prepare construction documents.

_____ 17. How well each refined idea meet the needs of the project is determined with a(n) _____ analysis.

_____ 18. True or False? A formal analysis of environmental impact is typically done during site analysis.

_____ 19. All of the following tasks may occur during structural analysis *except*:
   A. the strength of bridge designs are analyzed.
   B. the site is excavated.
   C. test borings are done.
   D. None of the above.

_____ 20. True or False? The drawings prepared for the client presentation can be used by contractors to build the project.

Name _____ Date _____ Class _____

Activity 11-1
Analyzing Commercial Architectural Drawings

Answer the following questions about a set of commercial architectural drawings provided by your teacher. Responsibilities should be shared and cooperation of the group members is essential to successful completion of this activity.

1. Find the title block on the first page of the drawings. It is typically located along the bottom edge of each page of the drawings.

   A. What is the title or name of the drawings?

   _____

   B. Who was the project designed for?

   _____

   C. Who prepared the drawings?

   _____

2. Find the plot plan.

   A. What is the scale of this drawing?

   _____

   B. If the property borders a street or road, what is the name of it?

   _____

   C. What is the maximum width and maximum depth of the property?

   _____

   D. What is the minimum distance from the street to the front of the building?

   _____

   E. What is the maximum width and depth of the building?

   _____

3. Find the first floor plan.

   A. What is the scale of this drawing? What fraction of actual size is this drawing?

   _____

   B. What is the largest room on the first floor?

   _____

Copyright by The Goodheart-Willcox Co., Inc.

C. How many restrooms are located on this floor?

D. How many staircases lead to other floors?

E. How many elevators are included in the plan?

4. Locate the front elevation of the building.

   A. How many floors are aboveground?

   B. What materials are visible on the front of the building?

   C. How many entrances are on this side of the building?

5. Find the first page of the foundation plan. Locate the walls around the perimeter of the building. Look for section lines along the perimeter walls like those shown below.

   Arrows indicates direction from which section drawing is viewed.

   Wall

   Location of section

   A. What letters or numbers identify the sections?

   B. Locate the page(s) that contains section drawings. Find the section drawing that corresponds to one of the label(s) found in Question 5A. What is the name of this drawing?

   C. What is the thickness of the foundation wall at this point?

   D. Is the wall supported with reinforcing rods?

Name _____

6. Continue using the foundation plan to answer the next two questions.

    A. Are footings or piers shown inside the perimeter walls? If yes, are they located in a grid pattern?
    _____

    B. What is the footing width and length?
    _____

7. Locate the structural drawings.

    A. What materials will be used to build the structure?
    _____

    B. What size columns are required for the first floor?
    _____

    C. Are the columns directly above the footings or piers?
    _____

    D. How long are the longest girders?
    _____

8. Locate the plumbing plan for the first floor. Identify the location of one drain-waste-vent stack (DWV) vertical pipe.

    A. What diameter is given for the DWV stack?
    _____

    B. In how many directions does horizontal pipe extend from the stack?
    _____

9. Locate the plumbing plan for the second floor.

    A. Is the DWV stack located directly above the DWV stack on the first floor?
    _____

    B. In how many directions does horizontal pipe extend from the stack?
    _____

10. Locate the heating, ventilating, and air-conditioning (HVAC) plans.

    A. Multistory buildings often have HVAC heating equipment on more than one floor. On what floors is heating equipment located?
    _____

    B. Where are the largest ducts installed on the first floor?
    _____

11. Locate the electrical plans.

   A. Where is the main electrical service panel located?
   _____

   B. Is there a secondary electrical service panel on the second floor? If yes, where is it located?
   _____

Chapter 12 Review
Construction Documentation

Name _____ Date _____ Class _____

Study Chapter 12 of the text, then answer the following questions.

Match each type of document with the correct description.

_____ 1. Estimate

_____ 2. Plan

_____ 3. Construction documents

_____ 4. Specifications

_____ 5. Working drawings

A. Papers that include working drawings, specifications, and job contracts.

B. Illustrations that describe the shape, size, and location of all parts of the project.

C. A drawing that shows a top view.

D. A part of the property that is set aside for utility installation and maintenance.

E. Description of the materials, construction methods, work quality that is required for a project.

F. A calculation of the cost to build a project.

_____ 6. True or False? Working drawings and specifications become a legally binding portion of the contract once the contract is signed.

_____ 7. A 1/2″ = 1′ scale drawing is _____ actual size.
   A. one-half
   B. one-twenty fourth
   C. one-sixth
   D. one-forty eighth

_____ 8. A plot plan shows the _____.
   A. size of the site
   B. placement and orientation of structures inside the property lines
   C. shape of the site
   D. All of the above.

Copyright by The Goodheart-Willcox Co., Inc.

_____ 9. True or False? Gazebos and other outdoor structures are shown on the foundation plan.

_____ 10. The base on which a foundation is built is called a(n) _____.

Match each type of drawings with the correct description.

_____ 11. Landscape plan

_____ 12. Elevation drawing

_____ 13. Site drawing

_____ 14. Electrical drawing

_____ 15. Mechanical drawing

_____ 16. Shop drawing

_____ 17. Structural plan

A. Shows the outside of a building as it would appear to someone looking straight at the building.

B. Shows elevator installation.

C. Shows the location of HVAC equipment, ducts, and controls.

D. Detailed drawings of parts to be made off site.

E. Shows the type and location of all wiring within a building.

F. Shows the type and placement of trees, shrubs, flowers, and water features.

G. Describes how the building is to be framed.

H. Depicts the site before construction begins.

_____ 18. Drawings that provide the information needed to build utilities, roads, pipelines, and other large projects are called _____ drawings.
   A. architectural
   B. structural
   C. engineering
   D. property

_____ 19. Specifications are used by _____.
   A. estimators
   B. material suppliers
   C. subcontractors
   D. All of the above.

_____ 20. Errors or omissions found in the specifications are corrected in the _____.

Chapter 13 Review
Project Management

Name _____ Date _____ Class _____

Study Chapter 13 of the text, then answer the following questions.

_____ 1. Accepted management practices include _____ activities.
  A. planning
  B. organizing
  C. controlling
  D. All of the above.

2. Identify each missing element in the management practices diagram.

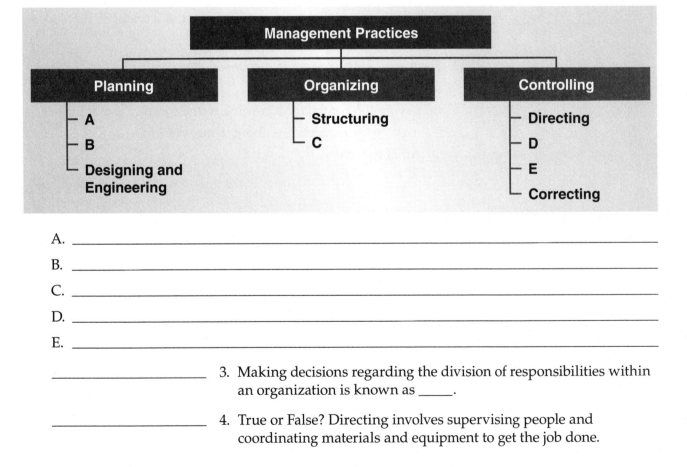

A. _____
B. _____
C. _____
D. _____
E. _____

_____ 3. Making decisions regarding the division of responsibilities within an organization is known as _____.

_____ 4. True or False? Directing involves supervising people and coordinating materials and equipment to get the job done.

Match each department with the correct description of work typically done by employees in that department.

_____ 5. Engineering department

_____ 6. Production department

_____ 7. Personnel department

_____ 8. Finance department

_____ 9. Marketing department

A. Keeps records of company income and expenses.

B. Implements strategies and policies that affect employees.

C. Advertises the services of the company.

D. Approves installation of mechanical equipment.

E. Responsible for the management of all work at the construction site.

10. Explain the difference between a project manager and a construction superintendent.

_____

_____

_____

_____

_____ 11. True or False? A job supervisor manages the work of all employees of a subcontractor working on a job site.

_____ 12. The owner of the project is responsible for _____.
   A. supplying the building site
   B. paying the contractor according to the contract terms
   C. making quick decisions about labor problems
   D. All of the above.

Name _____ Date _____ Class _____

Activity 13-1
Responsibilities of Contractors and Employees

In this activity, you will act as a successful general contractor and identify examples of management tasks you perform. You are a small custom homebuilder who builds an average of four custom homes each year. You also do a wide variety of maintenance and repair projects throughout your community. You are considering expanding your business to include multifamily housing.

You have eight tradespeople on the payroll who you want to retain long term. Overtime work for these employees can total up to 10 hours per week. You subcontract concrete, masonry, plumbing, HVAC, and electrical work. Your regular employees do all other tasks common to custom residential construction. You would like to add up to five temporary employees when the work load is heavy. You want to operate your company efficiently and build a strong reputation for building high quality sustainable homes.

Your office staff includes an assistant manager and an administrative assistant. The assistant manager helps you prepare estimates, but you finalize bids and negotiate contracts with clients and supervise all on-site work. The assistant manager also acquires materials, special tools and equipment, and maintains the company's financial records. The administrative assistant performs a variety of tasks, primarily to help you and the assistant manager.

You also pay a certified public accountant (CPA) to review the accounting practices and prepare the company's tax returns. You consult with legal counsel regarding contracts and legal requirements related to your business. Twice a year, you hire an advertising agency to prepare and distribute advertising for your company.

You are an active member of two civic associations and the local contractors association. You chair the local committee on sustainable construction certification.

1. Think about what management tasks you perform within the subcategories of planning, organizing, and controlling. Briefly describe three examples of tasks you perform other than those listed here and in the textbook.

   Planning: Formulating

   _____
   _____
   _____

   Planning: Researching

   _____
   _____
   _____

Copyright by The Goodheart-Willcox Co., Inc.

Planning: Designing and Engineering

_____
_____
_____

Organizing: Structuring

_____
_____
_____

Organizing: Supplying

_____
_____
_____

Controlling: Directing

_____
_____
_____

Controlling: Monitoring

_____
_____
_____

Controlling: Reporting

_____
_____
_____

Controlling: Correcting

_____
_____
_____

2. Review the description of your job as a homebuilder provided at the beginning of this activity. Which of your activities are related to advertising?

_____
_____
_____

Name _____

3. What are you doing to meet legal requirements?

   _____
   _____
   _____

4. Once you make contact with a potential client, what do you do to obtain a contract with the client?

   _____
   _____
   _____

5. What five characteristics, other than the knowledge and skill to do the work, would you want the tradespeople you employ to possess?

   _____
   _____
   _____
   _____
   _____

6. You may be asked to share some of your responses with the class.

Copyright by The Goodheart-Willcox Co., Inc.

Name _____ Date _____ Class _____

 # Activity 13-2
Green Construction Practices

The Green Construction feature in Chapter 13 addressed the role of management in sustainable construction. In this activity you will consider ways an individual worker can help reduce carbon emissions, improve performance of buildings, and reduce waste on the job site. Review this Green Construction feature before answering the following questions.

1. Identify three actions you can take as a construction worker to reduce carbon emissions.

   _____
   _____
   _____

2. Describe three actions you can take to reduce waste, promote recycling, and improve efficiency as you work on construction projects.

   _____
   _____
   _____

3. Identify three locations in your community that might use recycled and salvaged materials from remodeling and new construction projects.

   _____
   _____
   _____

4. Describe three actions you can take to minimize heat transfer through walls.

   _____
   _____
   _____

5. Identify three products used in construction that are made from at least 30 percent recycled materials. You may use the Internet to search for answers.

   _____
   _____
   _____

Copyright by The Goodheart-Willcox Co., Inc.

Chapter 14 Review
Construction Estimating and Bidding

Name _____ Date _____ Class _____

Study Chapter 14 of the text, then answer the following questions.

1. What is the purpose of an estimate?
   _____

2. Briefly explain the difference between the way a rough estimate and a detailed estimate are done.
   _____
   _____
   _____

_____ 3. Which of the following are *not* included in a detailed estimate?
   A. Tools and equipment.
   B. Materials and supplies.
   C. Labor.
   D. Profit.

4. The estimating process consists of specific steps done in order. Place the letters A–C in the blanks to indicate the order in which the steps are typically completed.

   _____ Visit the job site to collect information on site access and obstacles and availability of utilities.

   _____ Review working drawings and specifications.

   _____ Break the project into a sequence of work units.

_____ 5. A detailed list of materials for a construction project is called a(n) _____.
   A. bill of material
   B. takeoff
   C. materials list
   D. shipping list

Copyright by The Goodheart-Willcox Co., Inc.　　　　135

6. Compute the cost of 550 feet of electrical cable cost that sells for $0.35 per foot. Show all your work in a neat, orderly fashion.

7. A room measures 11' 9" × 16' 4". Calculate the cost of carpeting this room with carpet priced at $23.40 per square yard. Show all your work in a neat, orderly fashion.

8. Calculate the number of cubic yards of concrete required to build a retaining wall that is 9" thick by 6' 6" tall by 75' 3" long. Show all your work in a neat, orderly fashion.

_____ 9. True or False? Labor costs include wages, payroll tax, and worker's compensation insurance.

Name _____

Match each term with the correct description.

_____ 10. Depreciation

_____ 11. Overhead

_____ 12. Worker's compensation insurance

_____ 13. Interest

_____ 14. Markup

_____ 15. Profit

_____ 16. Loss

A. Unforeseen costs.

B. Loss in value of equipment as it gets older.

C. An amount added to an estimate to cover contingencies and to provide profit to the contractor.

D. Coverage that compensates employees who are injured in work-related accidents.

E. A situation in which expenses exceed income.

F. The ongoing expenses of running a company.

G. Money paid to a lender for the use of borrowed money.

H. The money that remains after all expenses are paid.

_____ 17. True or False? In competitive bidding, the owner and contractor meet to discuss estimated costs.

Chapter 15 Review
Construction Scheduling

Name _____ Date _____ Class _____

Study Chapter 15 of the text, then answer the following questions.

_____ 1. Which of the following statements regarding scheduling is true?
   A. A preliminary schedule may be prepared as part of the estimating process.
   B. Managers use the schedule to determine when workers, materials, and equipment will be needed.
   C. Schedules can be used to track monetary expenditures.
   D. All of the above.

_____ 2. The experience method of scheduling is best suited for _____ projects.

3. Bar charts are prepared in a particular sequence. Place the letters A–J in the blanks to indicate the order in which the steps are typically completed.

   _____ List tasks in sequence on bar chart

   _____ Calculate and enter percentage of job cost

   _____ Divide project into tasks

   _____ Draw horizontal bars representing time

   _____ Determine the number of tasks that can be overlapped

   _____ Obtain cost information for each task

   _____ Enter horizontal bars indicating when each task was done

   _____ Determine sequence of task completion

   _____ Enter cost data for each task in table

   _____ Estimate time to complete each task

_____ 4. An overall progress chart shows planned and actual _____ of money for a project.

Match each critical path scheduling method term with the correct symbol.

_____ 5. Critical path

_____ 6. Task

_____ 7. Event

A. ④

B. →

C. ⇒

_____ 8. Workers who are employed by one company on a consistent basis are known as _____ employees.
   A. illegal
   B. overtime
   C. regular
   D. temporary

9. Name two actions a contractor can take when the workload exceeds what can be completed in a standard workweek.

_____

_____

10. Identify five actions contractors can take when planning for safety.

_____

_____

_____

_____

_____

_____ 11. Material that is owned and stored by a construction company is known as _____ material.

_____ 12. A fixed amount of money that is available to the job superintendent for paying bills that are due immediately is called a(n) _____.

_____ 13. Contractors obtain equipment by
   A. purchasing or renting
   B. leasing
   C. contracting for the services of a subcontractor
   D. All of the above.

14. Name four types of permits that are needed when building a new house.

_____

_____

_____

_____

Name _____ Date _____ Class _____

Activity 15-1
The Critical Path Method

The critical path method (CPM) of scheduling uses a network to illustrate the relationship of the various tasks involved in completing a construction project. The network is composed of three major elements: activities, events, and the critical path.

Activities are represented by arrows and indicate activities or tasks to be completed. A brief description of the task is written above the arrow and the estimated time for completion is shown below the arrow.

Events are shown as numbered circles. They identify the beginning and end of activities. They are instants in time and therefore, do not require time estimates. The event numbers at the beginning and ending of each activity identify the activity, such as Activity 2-3.

The critical path is the path through the network that requires the greatest amount of time to complete. Arrows with double line shafts distinguish these activities from those not on the critical path. The total length of the estimated times along the critical path is the least amount of time in which the project can be completed.

Study the CPM schedule shown in the following figure. Then answer the following questions.

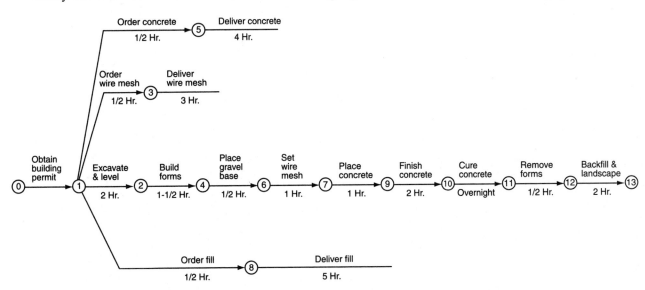

1. Add activity arrows to complete the paths for *Deliver concrete*, *Deliver wire mesh*, and *Deliver fill*.

_____

_____

Copyright by The Goodheart-Willcox Co., Inc.

2. Calculate the time required to complete each path through the network. What is the minimum work time in which the project can be completed?

___

___

3. Add a second line to each of the activity arrows along the critical path.

___

Chapter 16 Review
Site Preparation

Name _____ Date _____ Class _____

Study Chapter 16 of the text, then answer the following questions.

1. Name eight tasks that must be done to prepare a site before construction can begin.

   _____
   _____
   _____
   _____
   _____
   _____
   _____
   _____
   _____

_____ 2. A(n) _____ party consists of a party chief, the recorder, and instrument operator.

_____ 3. True or False? The party chief takes notes on the survey findings.

Match each term with the correct description.

_____ 4. Topographic survey

_____ 5. Transit

_____ 6. Sighting level

_____ 7. Markers

_____ 8. Surveyor's level

A. Objects that fix a location of the surface of the earth.

B. An examination of a parcel of land done to establish the land's contours and its features on or below the surface.

C. An instrument used to sight a horizontal line.

D. Survey method in which a line is established along the center of the site.

E. An instrument used to establish elevations and measure horizontal angles.

F. An instrument used to measure horizontal and vertical angles.

9. The surveying process involves five steps. Place the numbers 1–5 in the blanks preceding the five steps to indicate the sequence in which the steps are typically completed:

_____ Plotting—Place new markers to establish property boundaries.

_____ Research—Obtain information from previous surveys.

_____ Drawing—Create scale drawings of the site.

_____ Fieldwork—Locate previously installed markers.

_____ Purpose—Determine the reason for the survey.

_____ 10. A surveying method that uses a horizontal line of sight as a reference to measure elevation changes is called _____ leveling.

_____ 11. Which of the following are typically shown on topographic drawings?
   A. Contour lines
   B. Elevations
   C. Easements
   D. All of the above.

Match each term with the correct description.

_____ 12. Disposal

_____ 13. Salvaging

_____ 14. Blasting

_____ 15. Wrecking

_____ 16. Earthmoving

_____ 17. Baseline

_____ 18. Slope stake

_____ 19. Centerline

_____ 20. Batter boards

A. A process that changes the surface of a site by shifting the soil.

B. A series of reference points used to establish the position of a building on a site.

C. A demolition method that uses explosives to destroy structures and other objects.

D. Removing materials from buildings and other structures for use elsewhere.

E. Stake-and-board construction that marks the location of a structure.

F. The removal of debris from a site.

G. Reference points that pass through the middle of a project and indicate horizontal and vertical placement of the project.

H. A demolition method done using machinery.

I. An earthmoving process that removes high spots and fills low spots.

J. Markers that indicate the limit of the cut and fill at a given location.

21. Identify three utilities that are needed at a work site when construction begins.

_____

_____

_____

Chapter 16 Site Preparation 145

Name _____ Date _____ Class _____

Activity 16-1
Locating Property Boundaries

This activity simulates the process of locating property boundaries. You will work in small groups making measurements to lay out the boundaries. You will use a tape measure to measure distances and a direction circle to measure horizontal angles. Taking accurate measurements is key to successful completion of the activity.

Convert all length dimensions to the equivalent distance using a scale of 1/2" = 1'. Make angle measurements using the direction circle found at the end of this activity. The direction circle is a substitute for a surveyor's level or transit and uses a string to mimic the line of sight established with a surveyor's level or transit. Review the plot plan for the property, **Figure A**.

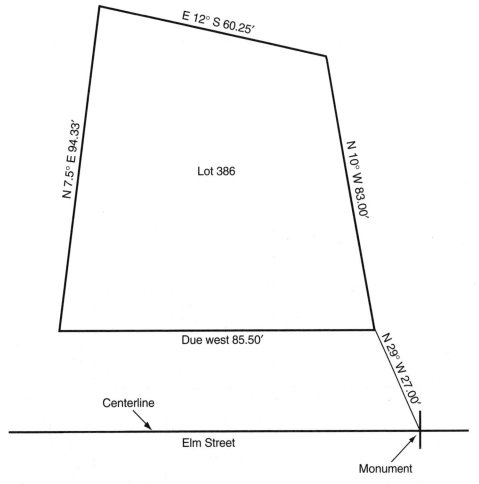

Figure A—Plot Plan

Copyright by The Goodheart-Willcox Co., Inc.

Layout is done on a 1/2" × 4' × 4' sheet of plywood. A separate notched board is used as monument simulator and is placed outside the plywood sheet. It is located at the centerline of the street. You will need the following tools and materials:

- 1/2" × 4' × 4' plywood sheet
- Monument simulator
- Hammer
- Framing square
- 12" rule
- Pencil
- Masking tape
- Eight 4d finish nails
- Four pieces of string, each 5' long
- One piece of string, 2' long

Procedures

1. Place the 4' × 4' sheet of plywood on a bench. Position the edge adjacent to the street parallel to one edge of the bench, **Figure B**.

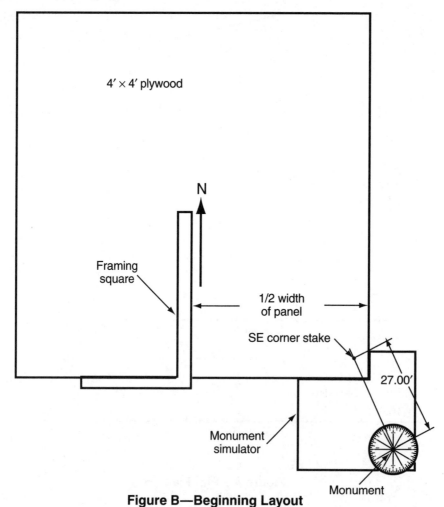

Figure B—Beginning Layout

Measure the width of the panel and locate the center. Use a framing square to draw the shaft of the north direction arrow. Add an arrowhead and the letter *N* to complete the direction arrow.

Name _____

2. Cut out four direction circles. Punch a small hole at the center of each circle using a 4d finish nail.
3. Position the monument simulator at the southeast corner of the plywood. Refer again to **Figure B**. Position a direction circle over the monument with a nail. Use the framing square to align the north-south axis parallel to the north direction arrow. Use masking tape to secure the direction circle to the monument simulator at 90° W and 90° E.
4. Attach the 2′ piece of string to the stake and stretch the string to align with N 25° W. Measure 27′ along the string. Drive a finish nail adjacent to the string and *not* more than halfway through the plywood. This nail indicates the southeast (SE) corner stake. Note: Dimensions used in surveying are written in this form: N 5° E 16.00′. This is read north 5 degrees east 16 feet. Letter abbreviations are N for north, S for south, E for east, and W for west. To measure an angle, position the direction circle so that its north-south axis is parallel with the North direction arrow. This simulates what a surveyor would do with the compass in the base of a surveyor's level or transit to align with the north and south poles of the earth. Stretch the 2′ string from the monument so that it crosses the direction circle 5 degrees west of due north. This establishes a line equivalent to the line of sight a surveyor would set up with a surveyor's transit. The final element of the dimension is the distance from a known point (the monument in this example) measured along the sighted line. Fractions of a foot are shown as decimals. For example, .75′ = 3/4′ = 9″.
5. Remove the string, direction circle, and monument simulator. Place the direction circle over the southeast corner stake with a nail. Use the framing square to align the north-south axis of the direction circle parallel to the north direction arrow, **Figure C**.

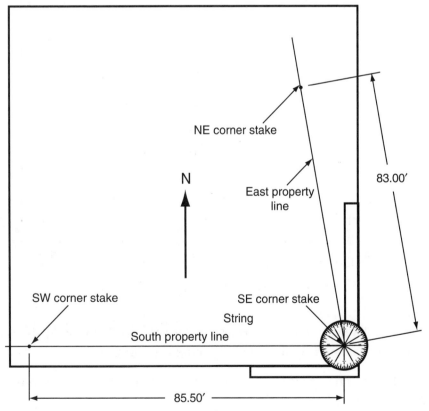

Figure C—Establishing East and South Property Lines

Use masking tape to secure the direction circle at 45° W and 45° E. Attach one 5′ length of string to the southeast corner stake. Stretch the string due west and measure 85.50′ along the simulated line of sight. This is the position of the southwest corner stake. Secure the end of the string to the southwest corner stake. The string represents the south property line.

6. Locate the northeast corner stake by measuring from the SE corner stake N 10° W 83.00'. Drive a finish nail adjacent to the string and *not* more than halfway through the plywood.

7. Position a direction circle at the SW corner stake and secure it with masking tape at N 30° E and S 40° E. Use the back-sighting process to position the northwest (NW) corner stake. This process is often used once two corner stakes are established to measure angles. The transit is set up at the second corner stake and focused on the first stake. The direction circle on the transit is adjusted to the desired angle as a basis for measuring the angle of the next boundary line. This sighting back to the previously established location is called back-sighting. Since the south property line in our example runs due east-west, the direction circle can be positioned so that the east aligns with the string marking the south property line. Tape the direction circle in this position and check the north-south axis using the framing square. Adjust and re-tape the direction circle as needed.

8. Attach a 5' or longer string to the SW corner stake and complete the installation of the NW corner stake N 7.5° E 94.33'. Secure the string to represent the west boundary of the property, **Figure D**.

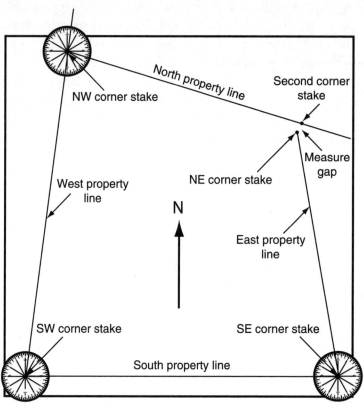

Figure D—Establishing West Property Line

9. Repeat Steps 7 and 8 from the NW corner stake E 12° S 60.25' to establish the north property line. Before driving the last nail, answer the following questions:

   A. If this nail is driven in the location measured, how far will it be from the previously installed NE corner stake?

   _____

   B. How far is this in actual distance?

   _____

Name _____

    C. If the distance measured in A is less than 1/2", attach the string to the previously established NE corner stake and proceed to Step 10. If the gap is greater than 1/2", continue with D.

    D. Recheck all distance and angle measurements. Begin with the SE corner stake and make any necessary adjustments. Continue in the same sequence used to establish the corner stakes.

    E. Measure the distance again between the two NE corner stakes. What is it?

    _____

    F. How far is this in actual distance?

    _____

    G. If the distance between the two northeast corner stakes is less than 1/4", secure the string to the first northeast corner stake. If the distance is greater than 1/4", drive the second corner stake at the measured location at the end of the north boundary line. Secure the string.

10. Staple these pages together, place your name on the first page, and turn the activity in to your instructor.

11. Return tools and materials to storage.

Name _____

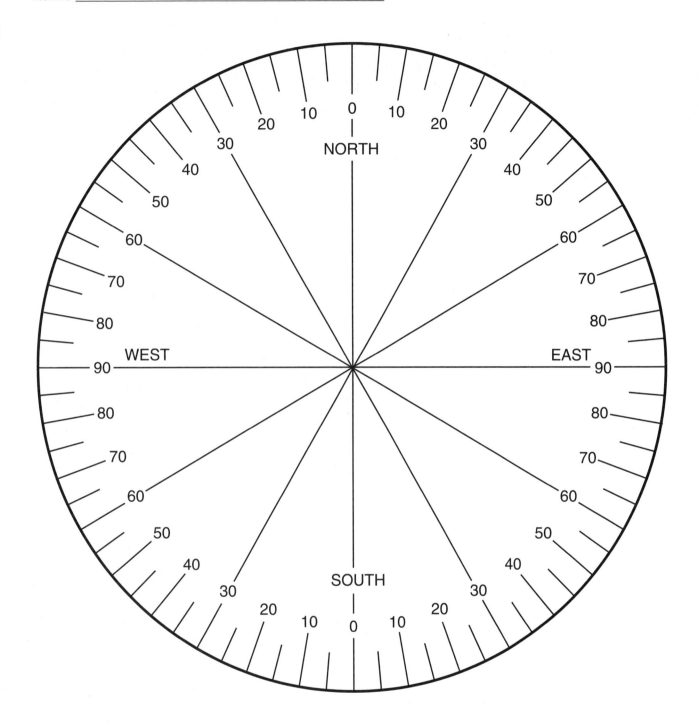

Name _____ Date _____ Class _____

Activity 16-2
Locating Structures

Accurately locating the four corners of a structure on a construction site is essential to any new project. Once the position of the four corners is established, batter boards are constructed. These boards are placed far enough outside the excavation area that they are not disturbed during excavation. This allows strings to be repositioned marking the location of the structure without repeating the layout process.

Your instructor will tell you where to begin the layout for a 8′ × 12′ storage building. This location will be an existing sidewalk, building, curb, or other stationary object. You will lay out the building parallel to this object. When finished, your layout will look like this:

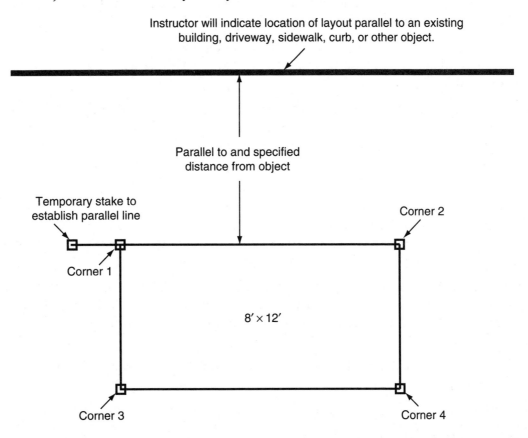

Based on the information provided by your instructor, complete the following sentence:

- The 12′ side of the storage building should be parallel to and _____ feet away from the _____.

Along with this information, you will need the following tools and materials:

- Five 2 × 2 stakes
- Transit and tripod
- 15′ or longer tape measure

Copyright by The Goodheart-Willcox Co., Inc.

- One 3 pound hammer
- One nail hammer
- Four 6d finish nails

1. Set up and level the transit approximately 20' away from the nearest side of the storage building site.

2. Measure the distance between the storage building location and the parallel object. Drive one stake into the ground at this point, leaving the top at least 12" aboveground. Remeasure the distance between the stake and object and adjust the distance as needed.

3. Locate the position of a second stake the same distance from the object and at least 14' away from the first stake. Drive this stake far enough into the ground for it to be stable. Measure the distance from the object to the stake and adjust as needed.

4. Use the transit to check the elevation of the top of the two stakes. Drive the higher stake farther into the ground until the tops of the two stakes are at the same elevation.

5. Drive a nail into the top of each stake at the specified distance from the parallel object. This measurement needs to be accurate. Hold the tape measure so it is level, at a 90° angle to the parallel object, and with as little sag in the middle as possible.

6. Attach a string to the nail in the top of one of the stakes. Stretch the string and secure it to the nail in the second stake. Measure 12' from the second stake along the string and drive a stake that is centered on the string line.

7. Check the elevation of the third stake. Adjust the height until it is equal to the first stake. Check the position of the string line to locate center of the stake. Again, measure 12' from the nail in the top of the first stake and drive a nail in the top of the third stake, alongside the string and 12' from the first nail. Remove the string from the first stake and secure it to the nail in the third stake. The first two corners of the storage structure are now located.

8. Reposition and level the transit over the first stake such that the plumb bob is directly above the nail. Rotate the scope of the transit both horizontally and vertically and focus on the nail in the top of the stake at the second corner of the storage building. Rotate the direction circle at the transit base to align 0° with the pointer and lock the direction circle.

9. Rotate the scope 90° and lock the horizontal rotation. Measure 8' from the first stake. Rotate the scope vertically and move the tape measure sideways until the scope is focused on the 8' mark on the tape measure. Drive a stake at this point until the top is the same elevation as the top of stake 2.

10. Check the 8' measurement and the angle. Make any adjustments and drive a nail in the top of the third stake to locate corner 3.

11. Reposition and level the transit over the corner 2 stake and repeat Steps 8–10 to locate corner 4.

12. Answer the following questions:

    A. Do the two longer sides measure 12'?

    _____

    B. Do the two shorter sides measure 8'?

    _____

Name _____

C. What is the distance between corner 1 and corner 4?

_____

D. What is the distance between corner 2 and corner 3?

_____

E. If the distances measured in C and D are not equal, one or more of the corners is not square. Review your answers to A–D and explain what must be done to correct the problem.

_____

_____

F. Reposition and level the transit 15′ to 20′ away from the 8′ × 12′ rectangle. Lock the scope in the horizontal position. Complete rows 1 and 2 in the following table.

Row	Distance from:	Corner 1	Corner 2	Corner 3	Corner 4
1	Line of sight to top of stake				
2	Line of sight to ground at base of stake				
3	Top of stake to ground				
4	Top of fill ground				

G. Calculate the distance from the top of each stake to the ground at the base of each stake and complete row 3 in the table.

H. Fill will be needed to create a level building site for the storage building. Calculate the amount of fill needed at the base of each stake to produce a site that is level with the ground at the highest corner of the site. Record the results of your calculations in row 4.

13. Have your instructor evaluate your work. If time permits, return to Step 12 and correct any problems with the layout. Otherwise, remove the stakes and put away all tools and materials. Fill the holes made by the stakes.

Chapter 17 Review
Earthwork

Name _____ Date _____ Class _____

Study Chapter 17 of the text, then answer the following questions.

_____ 1. Soil is generally made up of varying amounts of minerals, air, water, and _____ matter.

_____ 2. Changes made to the earth on a construction site are known as _____.

3. Briefly explain the difference between bank soil and compacted soil.

_____

_____

_____

_____ 4. To prevent cave-ins during or following excavation, builders will _____.
   A. slope the sides of the excavation
   B. build retaining walls
   C. treat the soil
   D. All of the above.

5. Identify the parts of this retaining wall.
   A. _____
   B. _____
   C. _____

_____ 6. True or False? Cofferdams support trench walls.

7. Name four methods for loosening soil.

_____
_____

_____ 8. Which of the following statements regarding excavation is *not* true?
   A. It involves loosening dense soil through the use of vibration.
   B. Is used to create a strong surface to support the structure being built.
   C. Is the process of digging, moving, and removing earth at construction sites.
   D. Is done using hand tools or machines.

9. Briefly explain why foundations must extend below the frost line.

_____
_____

Match each type of excavation with the correct description.

_____ 10. Bulk-pit excavation
_____ 11. Bulk wide-area excavation
_____ 12. Loose-bulk excavation
_____ 13. Limited-area vertical excavation
_____ 14. Trench excavation
_____ 15. Dredging

A. Moving loose material from one location to another without hauling.
B. Digging a small area to a considerable depth.
C. Removing soil or other material from below water.
D. Digging a wide, deep area and removing the spoil.
E. Boring small diameter horizontal holes through the earth.
F. Digging long, narrow holes.
G. Shallow digging done over a large, easily accessed area.

_____ 16. What type of excavation is done completely underground and requires the removal of all spoil?

17. List three machines that can be used to transfer spoil.

_____
_____
_____

_____ 18. Smoothing and leveling soils in preparation for landscaping or roadbeds is known as _____.
   A. stockpiling
   B. grading
   C. stabilizing
   D. breaking

Name _____ Date _____ Class _____

Activity 17-1
Selecting Earthmoving Equipment

Earthmoving equipment is available in many types and sizes. In this activity you will learn what types of equipment are used for different projects.

In the table below, the first column lists various construction projects. The second column notes the size of each excavation. In the third column, you will list several types of earthmoving equipment that might be used to perform the work. In those rows where you have listed more than one piece of equipment, circle the piece you think is the best choice for the project.

You will complete several rows together as a class. Complete the remainder of the table on your own outside of class.

Types of Project	Minimum Size of Excavation	Equipment Choices
Gas supply from main to building, 3/4" polyethylene	2" wide × 3' deep × 50' long	
Spread footings for exterior walls of house	2' wide × 30" deep × 130' long perimeter trench	
Basement for a home	7' 6" deep × 30' wide × 46' long	
Water main	2' wide × 4' deep × 1000' long	

(Continued)

Copyright by The Goodheart-Willcox Co., Inc.

Types of Project	Minimum Size of Excavation	Equipment Choices
Leveling gently sloped site for a house	60′ × 75′, average depth 6″	
Cutting through hill to build road	50′ wide × 450′ long, average depth 18″	
Clearing a shipping channel for ships	200′ wide × 2500′ long, average depth 4′	
Replacing water supply from main to building below driveway and large tree	1 1/2″ diameter × 75′ long	
Rock from side of mountain for roadway	1000 cubic yards	

Chapter 18 Review
Foundations

Name _____ Date _____ Class _____

Study Chapter 18 of the text, then answer the following questions.

1. Define the term *foundation*.
   _____
   _____

   _____ 2. True or False? The bearing surface is the point of contact between the substructure and the earth.

3. What is the purpose of footings?
   _____
   _____
   _____

Match each type of foundation with the correct drawing.

_____ 4. Spread foundation

_____ 5. Floating foundation

_____ 6. Friction pile

_____ 7. Bearing pile

_____ 8. Pier foundation

A.

B.

C.

D.

E.

_____ 9. True or False? Foundation walls are constructed before footings.

_____ 10. A(n) _____ is a groove molded down the center of a footing that strengthens the joint between the footing and the foundation wall.

Name _____

11. Name two materials used to make foundation walls.

_____
_____

12. Identify the parts of the wood foundation footing.

_____ Double plate
_____ Polyethylene sheet
_____ Concrete floor
_____ Plywood sheathing
_____ Foundation drain
_____ Floor framing
_____ Top plate
_____ Gravel fill
_____ Bottom plate
_____ Stud
_____ Insulation
_____ Footing

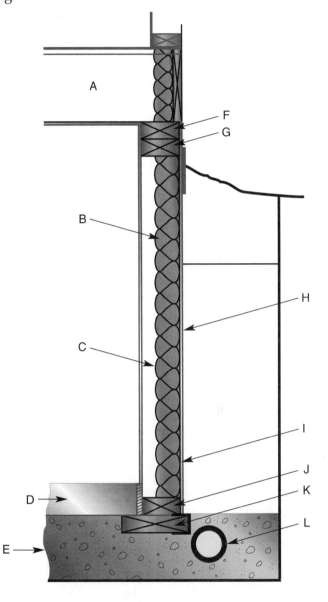

13. Briefly describe six things that can be done to increase the water resistance of a foundation wall.

_____
_____
_____
_____
_____
_____

_____ 14. The primary benefit to insulating foundations is _____.
　　　　　　　　　　　　　A. improved waterproofing
　　　　　　　　　　　　　B. reduced initial cost
　　　　　　　　　　　　　C. improved appearance
　　　　　　　　　　　　　D. reduced heating cost

Name _____ Date _____ Class _____

Activity 18-1
Green Foundation Materials

This activity focuses on green construction products used to build foundations for residential structures. With the help of your instructor, select one of the following green foundation materials. Prepare a report, presentation, or display that describes the product and focuses on its strengths and weaknesses compared to more traditional materials typically used for the same purpose in residential construction.

Possible topics include:

- Pressure-treated wood foundations.
- Fly-ash substituted for some of the Portland cement.
- Insulated concrete forms.
- Faswall® blocks.
- Rastra® blocks.
- Other topics as approved by your instructor.

Your instructor will provide specifics regarding your report such as length, presentation time, and display size.

Name _____ Date _____ Class _____

Activity 18-2
Concrete Activity

Complete Chapter 18 before beginning this activity. Part of this assignment requires working cooperatively with classmates.

Objectives

Upon completion of this laboratory assignment, you should be able to do the following:

- Properly treat forms prior to placing concrete.
- Properly mix a 1–2 1/4–3 batch of concrete in the portable concrete mixer.
- Slump test the batch of concrete, and correct the batch to the proper proportions.
- Properly place concrete in forms, and consolidate and finish the concrete.
- Allow the concrete to cure in the form prior to stripping or removing the form.
- Properly strip or remove and clean the form.
- Maintain a clean orderly work area, properly clean and store tools and machines, and return unused materials to storage.

Tools and Equipment

- One preassembled form for patio block
- One wood float
- One edger
- One straight edge made of a scrap piece of 2 × 4
- One portable concrete mixer or mortar box
- One shovel
- One tamping rod 5/8" diameter × 24" (reinforcing bar)
- One slump test cone
- One ruler
- One water hose
- One 6mm × 4′ × 4′ plastic sheet
- One wheelbarrow
- One broom

Copyright by The Goodheart-Willcox Co., Inc.

Materials and Supplies

- Portland cement
- Sand
- Gravel
- Oil for treating forms
- Water

Safety Notes

- Use a wheelbarrow to move bags of Portland cement, sand, and gravel. These items are heavy and can be difficult to grasp.
- Never put your hands or any tools inside a concrete mixer while it is rotating. The rotating paddles are dangerous.
- Both wet and dry Portland cement are harmful to eyes and skin. Wear safety glasses or goggles to protect your eyes. Wash exposed skin with soap and water as soon as possible.
- Clean up water and wet concrete spills immediately. Use sawdust or jointer chips to absorb the moisture.

Procedural Notes

Read the procedures and carefully examine all diagrams before class. Research any unfamiliar terminology and technical information so you are prepared to go to work in the laboratory.

Read each step again, carefully. As you complete each step, make a check mark in the blank that precedes each step. Steps preceded by a circle require your instructor's approval before continuing with the next step. If you do not obtain your instructor's approval, you will not receive full credit for the activity.

Group Activity—Mixing and Testing Concrete

Your instructor will divide you into groups of 3–5 people. Collect tools and materials, set up equipment, and divide tasks between the members of the group.

_____ 1. Gather all equipment, tools, and materials to the work area. Record your form number here: _____.

_____ 2. Treat your form with oil.

_____ 3. Calculate the volume of your form in cubic inches using this formula: thickness in inches × width in inches × length in inches = _____ cubic inches

_____ 4. One cubic foot is equal to 1728 cubic inches. Calculate the number of cubic feet needed for your block: _____ cubic feet. Calculate to two decimal places. Add to find the total needed by your group: _____ cubic feet.

Name _____

_____ 5. Determine the moisture content of the sand. Do this by squeezing a handful of sand together. Water is added to the sand based on your assessment of its moisture content. The amounts added are per cubic foot of concrete. The following table lists the amount of water to add:

Moisture Content of Sand	Attributes	Gallons of Water per Cubic Feet of Concrete
Dry	Does not compact when squeezed in the hand	1.4
Damp	Falls apart in the hand	1.3
Wet	Forms into a ball in the hand	1.2
Very wet	Sparkles and wets the hand when squeezed	1

The sand is _____, requiring _____ gallons of water per cubic foot of concrete.

_____ 6. Your mix will consist of one part Portland cement, two and one-quarter parts sand, and three parts gravel, plus enough water to produce a batch with the desired slump. Use the following formula to calculate the amount of water needed:

gallons of water per cubic foot of concrete × cubic feet of concrete = gallons of water needed

Measuring the water will be easier if gallons of water are converted to pints of water using the following formula:

gallons of water × 8 pints per gallon = pints of water needed

_____ 7. One cubic foot of concrete is made by combining 400 cubic inches of cement, 900 cubic inches of sand, and 1200 cubic inches of gravel. To calculate the volume of cement, sand, and gravel your group will need to mix, use the formulas given below. The additional 10 percent is for waste.

400 cubic inches cement × _____ cubic feet needed = _____ cubic inches + 10 percent
= _____ cubic inches cement

900 cubic inches sand × _____ cubic feet needed = _____ cubic inches + 10 percent
= _____ cubic inches sand

1200 cubic inches gravel × _____ cubic feet needed = _____ cubic inches + 10 percent
= _____ cubic inches gravel

_____ 8. Measure Portland cement, sand, and gravel using two sizes of scoops:

Large scoop = 1 gallon = 231 cubic inches
Small scoop = 1 pint = 29 cubic inches

170   Construction and Building Technology Tech Lab Workbook

○ 9. With this in mind and using the volumes calculated above, complete the following summary chart and have your instructor check your calculations.

Water	Cement	Sand	Gravel
_____ gallons	_____ cubic inches	_____ cubic inches	_____ cubic inches
_____ pints	_____ full gallons plus _____ pints	_____ full gallons plus _____ pints	_____ full gallons plus _____ pints

If you are using a concrete mixer, go to Step 10. If you are using a wheelbarrow or mortar box, go to Step 11.

_____ 10. Mixing with concrete mixer:

   A. Pour approximately 25 percent of the water required into the mixer. Set the drum in motion. Never put your hands or any tool into the drum while it is in motion.

   B. Gradually add all the sand, gravel, and Portland cement. Add all but 10 percent of the water.

   C. Run the mixer for 1 1/2 minutes after all dry materials is placed in the drum. Stop the drum and check the consistency of the mixture.

   D. If the mixture is not uniformly moist or is extremely stiff, add a portion or all of the reserved water.

   E. If more water was added, run the mixer one more minute.

   F. While the mixer is running, set the slump cone on a plastic covered surface near the mixer, narrow opening up. Continue with Step 12.

_____ 11. Mixing in a mortar box or wheelbarrow:

   A. Place the sand, gravel, and Portland cement in the mortar box or wheelbarrow.

   B. Thoroughly mix the dry ingredients with a shovel or mortar hoe.

   C. Add all but 10 percent of the water and mix until consistency is uniform.

   D. If the mixture is not uniformly moist or is extremely stiff, add a portion or all of the remaining water.

   E. If water was added, mix until concrete is uniformly moist.

   F. While the mixing is in progress, set the slump cone on a plastic covered surface near the mixer, narrow opening up.

_____ 12. Fill one-third of the cone with concrete and rod the concrete exactly 25 times.

_____ 13. Fill the second one-third of the cone and rod exactly 25 times. Rod deep enough to pass through the second one-third of the mix into the first one-third of the mix.

_____ 14. Add more concrete in the cone, filling it to the top. Rake off level and rod again 25 times, extending the rod through the top one-third into the middle one-third of the mix. Add enough concrete to completely fill the slump cone and strike the top surface.

Copyright by The Goodheart-Willcox Co., Inc.

Chapter 18 Foundations

Name _____

____ 15. Carefully lift the cone straight up to expose the fresh concrete. Place the cone next to the slump of concrete. Place the rod flat across the top of the cone so it extends over the slump. Measure from the top of the slump to the bottom edge of the rod.

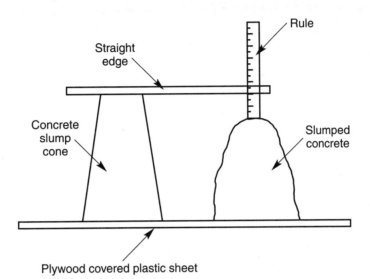

The distance between the rod and the slump is _____ inches.

____ 16. The slump must be no more than one-third the cone height, or 4 inches, and no less than 1/6 the cone height, or 2 inches. If the mix has excess slump, add cement and aggregate and retest. If the slump is insufficient, add water.

____ 17. Return the concrete from the slump test into the mixer and run it. Clean the cone and work area.

Individual Activity—Placing and Finishing Concrete _____

____ 18. Place a plastic sheet over the floor and beneath the form. Covering the work area with newspaper will make cleanup easier. Continuously place concrete from the mixer, mortar box, or wheel barrow into your form to a level slightly above the top of the form.

____ 19. Use the float to spread the concrete and compact it. Use a scrap stick to work the concrete along the edge of the form.

____ 20. Use a scrap 2 × 4 to strike off the concrete. Make a sawing motion back and forth across the surface. Some excess concrete should move in front of the straight edge to fill voids or valleys.

____ 21. Use the float to smooth the surface of the concrete until all large aggregate disappears from the surface and a layer of water appears. Allow your concrete to set and bleed.

____ 22. Clean the area and all tools except the float and edge. Clean the mixer and return unused cement, sand, and gravel to storage.

____ 23. After your concrete has set, use the edger to shape the outside edge of your patio block and float the surface as required.

_____ 24. After the water has begun to evaporate, trowel the surface for a smooth hard finish. Use the edger again if troweling disfigured the edging. Clean these tools.

◯ 25. Have your instructor inspect your block, work area, tools, and the mixer.

◯ 26. Allow your block to set overnight at the minimum. Then carefully disassemble the form and remove the block. Clean the form and reassemble.

◯ 27. Complete the student column of the Concrete Activity Evaluation. Then ask your instructor to inspect your work and complete the instructor's column.

Name _____ Date _____ Class _____

Activity 18-3
Concrete Activity Evaluation

Read each statement carefully. Decide how many of the possible points you earned and record this number in the student column. Then ask your instructor to evaluate your work.

Evaluation Standards	Maximum Points	Student	Instructor
1. Were the calculations of cement, sand, gravel, and water correct? (Step 9)	6		
2. Was the concrete mixed according to established procedure?	4		
3. Did the slump test meet established criteria?	5		
4. Were forms properly treated?	4		
5. Was the concrete properly consolidated?	4		
6. Is the surface true?	5		
7. Were the finishing and edging completed properly?	4		
8. Was the concrete allowed to cure properly prior to stripping or removing the form?	4		
9. Was the form cleaned properly after use? (Step 26)	5		
10. Were safe work practices followed?	4		
11. Was the work area clean and orderly? Were tools, equipment, and machines cleaned properly upon completion of the activity? Were extra materials returned to storage? (Step 26)	5		
Total Points	50		

Copyright by The Goodheart-Willcox Co., Inc.

Chapter 19 Review
Floors

Name _____ Date _____ Class _____

Study Chapter 19 of the text, then answer the following questions.

1. What is a superstructure?
   _____
   _____

2. Name four types of superstructure.
   _____
   _____
   _____
   _____

_____ 3. True or False? Most modern buildings have a frame superstructure.

_____ 4. Which of the following is *not* a type of frame superstructure?
   A. Wood
   B. Fabric
   C. Steel
   D. Reinforced concrete

_____ 5. A framing system in which the first floor is built on top of the foundation wall is known as _____ framing.

6. Identify the parts of the floor frame in the following drawing.

   A. _____
   B. _____
   C. _____
   D. _____
   E. _____

Copyright by The Goodheart-Willcox Co., Inc.

175

7. Buildings wider than 25 feet must have a(n) _____ approximately one-half the distance between the outside walls.
   A. steel I-beam
   B. wood girder
   C. bearing wall
   D. All of the above.

8. Place the numbers 1–9 in the blanks preceding the following steps to indicate the sequence in which the steps are typically completed.

   _____ Measure from the face of the foundation wall to the center of the anchor bolt and transfer this measurement to locate the center of the hole for the anchor bolt.

   _____ Removed the sill and install sill sealer.

   _____ Place the sill on top of the foundation wall, alongside the anchor bolts.

   _____ Drill the holes and put the first section of sill in place.

   _____ Reinstall and secure the sill with flat washers and nuts.

   _____ Straighten crooked anchor bolts.

   _____ Draw lines on each side of the anchor bolt, square with the edge of the sill.

   _____ Repeat the process until all of the sill is in place.

   _____ Check the fit and make needed adjustments.

9. The distance between vertical supports is called the _____.

10. The maximum allowable span for dimension lumber depends on the _____.
    A. size of the joists
    B. spacing of the joists
    C. strength of the joist lumber
    D. All of the above.

11. True or False? Joists must support both the weight of the structure and the weight within the structure.

12. Explain the difference between dead load and live load.

   _____
   _____
   _____

13. When framing openings in a floor frame, _____ run parallel to joists and _____ run perpendicular to the ends of the joists.

   _____

14. The curvature or bend along the edge of a board is known as the _____.

Name _____

15. Identify the framing members in the staircase opening shown.

    A. _____
    B. _____
    C. _____
    D. _____
    E. _____

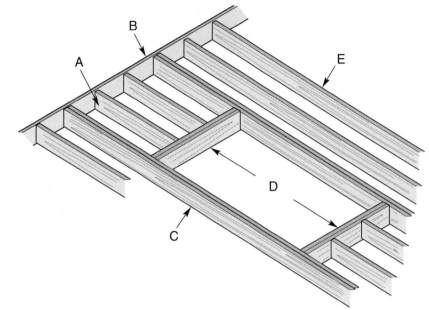

_____ 16. A support system that holds floor joists in a vertical position and distributes live loads to three or more floor joists is called _____.
   A. bridging
   B. heading
   C. casting
   D. spanning

17. What are the most commonly used subflooring materials?
_____
_____

_____ 18. True or False? The first step in subfloor installation is to lay out a chalk line 8′ from the outside edge of the joist header.

Name _____ Date _____ Class _____

Activity 19-1
Green Framing Materials

This assignment focuses on green construction products used in framing residential structures. With your instructor's help, select one of the following topics about green framing materials. Prepare a report, presentation, or display that describes the product and focuses on its strengths and weaknesses compared to materials traditionally used for the same purpose in residential construction.

Possible topics include:

- Sustainably harvested lumber.
- I-joists.
- Parallel stand lumber.
- Open-web floor joists.
- Laminated-veneer lumber.
- Glue-laminated beams.
- Oriented strand board.
- Rigid foam wall sheathing.
- Finger-jointed lumber.
- Structural insulated panels.
- Agriboard products.
- Homasote structural products.
- Other topics as approved by your instructor.

Your instructor will provide specific directions for your report regarding things such as length, presentation time, or display size.

Copyright by The Goodheart-Willcox Co., Inc.

Name _____ Date _____ Class _____

Activity 19-2
Floor Joist Layout

In this activity you will practice laying out full-size floor joists using common tools. Spacing floor joists is critical because 4' × 8' plywood or OSB sheets are used for subflooring. The ends of these sheets must meet at the center of a floor joist so they can be nailed properly. Since the subfloor must extend all the way to the edge of the floor frame, an adjustment is made in the normal 16" spacing of the first joist from the end of the floor frame. The adjustment is one-half the thickness of the floor joist. Locating the remaining joist 16" on center from the second joist ensures that the 4' × 8' sheets of subflooring will end at the center of a floor joist.

You will need the following tools and supplies:

- Tape measure
- Try square or framing square
- 10' 2 × 4 or 3 1/2" wide × 10' long strip of paper
- Pencil
- Eraser

1. Use the 2 × 4 or the strip of paper to lay out seven floor joists spaced 16" on center. The first and second sheets of 4' × 8' subflooring must butt together at the center of the seventh joist. Assume that the joists are nominal 2" lumber. Mark the location of each joist with an X beside the line, indicating one edge of the joist.

2. Have your instructor check your work.

3. A 3' 4" wide stairwell opening will be framed into the floor. Center this opening 2' 9" from the end of the floor frame. Modify the above layout by marking the location of stringers with an S and draw a line through joist locations that will not have full-length floor joist.

4. Add double joist outside the two stringers.

5. Have your instructor check your work.

Chapter 19 Floors 183

Name _____ Date _____ Class _____

Activity 19-3
Drawing Floor Frames

Working through the process will help you understand how floor frames are assembled. In this activity, you will draw a floor frame for a structure. Draw a plan view, showing the top edges of the floor joists and headers, and the face of the sills that are below the floor joists.

You will need a pencil and eraser, in addition to the graph paper included in this activity.

1. Determine the overall dimensions of the floor frame by referring to the drawings for the assigned structure. What is the overall width and length of the floor frame?

   _____

2. Locate the graph paper labeled "Floor Frame" at the end of this activity. Determine the size of the squares. How many squares per inch of length?

   _____

   If the scale of the drawing is 1" = 1', what distance is represented by one square?

   _____

   Will the floor frame drawing fit within the borders and allow space for dimensions? If not, can you change the scale so the drawing will fit? What scale will you use?

   _____

3. Drawing lightly, lay out a rectangle equal in size to the width and length of the required floor frame. Center the rectangle within the borders of the graph paper.

4. Check the assigned drawings to determine the location and size of the sills. Lightly draw the inside edges of the sills parallel to the correct sides of the floor frame.

5. Assume that 4' × 8' subflooring will be used. Lay out the position of each floor joist. Nominal 2" lumber is to be used for floor joists and headers. Lightly draw the top edges of the floor joists.

6. Draw the top edges of the headers at each end of the floor frame.

7. Erase overlapping and unnecessary lines and darken the object lines of the floor frame.

8. Add dimensions for floor joists and the overall floor frame. Label at least one header, one sill, and one floor joist.

9. Complete the title block and submit your drawing and this activity for grading.

Copyright by The Goodheart-Willcox Co., Inc.

Name _____

Drawn by:	Date:	Drawing:	Project:
		Floor Frame	

Name _____ Date _____ Class _____

Activity 19-4
Observation Form—Portable Electric Circular Saw

Read this observation entirely before working with your learner. Then follow each step as you conduct the practice session for the learner. As you complete each step, make a checkmark in the blank that precedes each step.

_____ 1. Set up the tools and supplies listed below in the work area:
- Portable electric circular saw, unplugged, blade removed.
- Extension cord
- Sawhorse
- Try-square
- 1 × 6 approximately 2′ in length
- 2 × 4 approximately 2′ in length
- Oil can
- Rag
- Wrench for blade installation and removal

_____ 2. Ask your learner to install the blade on the saw. Observe the procedures listed in the table while the blade is being installed. Record your observations. Refrain from giving any assistance until the assembly is complete or the learner is unable to proceed.

Procedure	Yes	No
A. Teeth of blade pointing in the correct direction.		
B. Oil on D-ring washer.		
C. Nut or bolt holding blade securely tightened.		

_____ 3. If there are any errors, ask the learner to correct the errors.

_____ 4. Ask the learner to mark the 1 × 6 (or 2 × 4 if this is a second try) for a square cut approximately 2″ from one end. They should not mark through knots or similar obstructions.

Copyright by The Goodheart-Willcox Co., Inc.

_____ 5. Direct the learner to cut the board so that the kerf is on the waste side of the line (the shorter piece of stock). Make the following observations as the crosscutting operation is being performed. Record your assessments. Stop the operation immediately if you see any safety issues or other problems. Call these issues to the attention of the learner in a way that will not distract or surprise the learner from the work being done.

Procedure	First Try		Second Try	
	Yes	No	Yes	No
A. Loose clothing removed or secured.				
B. Eye protection used without being told.				
C. Depth adjustment made correctly.				
D. Board solidly supported on top of, not at right angle to, top of sawhorse.				

Do not plug the saw in until A–D are correctly done.

Procedure	First Try		Second Try	
	Yes	No	Yes	No
E. Feet solidly placed about shoulder width apart.				
F. Body position such that elbow and shoulder are aligned with plane of saw cut.				
G. Saw allowed to come to full speed before cut begins.				
H. Rate of feed too slow.				
I. Rate of feed too fast.				
J. Saw permitted to stop rotating before being set down.				
K. Saw turned off.				
L. Saw set down to avoid resting on retractable blade guard.				

Name _____

_____ 6. Using the try square, test the squareness of the cut and record the results in the following table.

Cut	First Try 1 × 6	Second Try 2 × 4
A. More that 1/8" out of square		
B. 1/16" to 1/8" out of square		
C. Less than 1/16" out of square		

_____ 7. Review your ratings recorded in Step 6 with your learner.

_____ 8. If errors were made, be sure your learner understands what they were and how to correct them. Then ask your learner to repeat Steps 4–7. Record your evaluations in the appropriate location in the tables in Steps 5 and 6.

_____ 9. Ask your learner to sign and date this form.

I have completed this circular saw practice and reviewed the evaluation with my student trainer.

Learner _____ Date _____

Chapter 20 Review
Walls

Name _____ Date _____ Class _____

Study Chapter 20 of the text, then answer the following questions.

_____ 1. True or False? Curtain walls enclose a structure but do not provide structural support.

2. Identify the parts of the platform frame wall.

   A. _____
   B. _____
   C. _____
   D. _____
   E. _____
   F. _____
   G. _____
   H. _____

_____ 3. In platform framing _____.
   A. the walls are framed before the floor platform is in place
   B. the floor platform is in place before the walls are framed
   C. the walls and floor platform are assembled at the same time
   D. None of the above.

4. What is the first step in laying out a wall frame?

   _____
   _____

_____ 5. Standard stud spacing is _____ inches on center.
   A. 16
   B. 18
   C. 24
   D. 16 or 24

Copyright by The Goodheart-Willcox Co., Inc.

_____ 6. The older, nonliving center layers of a tree are called _____.

7. Rough openings provide spaces for _____ and _____.
_____

8. List three safety tips to keep in mind when working with a circular saw.
_____
_____
_____

9. What does a wavy line indicate when it is drawn between two trimmers?
_____
_____

10. What is the purpose of additional studs and blocking installed at corners and T-intersections?
_____
_____
_____

_____ 11. True or False? Headers for door and window openings are usually the same height.

_____ 12. Use _____ bracing when sheathing materials are not very rigid.
    A. no
    B. diagonal
    C. horizontal
    D. vertical

13. Briefly describe two innovations in wall framing.
_____
_____
_____
_____

Name _____ Date _____ Class _____

Activity 20-1
Drawing a Wall Frame

Draw the framing for one wall of a module, storage building, or other structure as assigned by your instructor. Your drawing should show the edges of plates, studs, cripple studs, rough sills, and the face of any headers.

You will need a ruler, pencil, eraser, and the graph paper included at the end of this activity.

1. Determine the overall dimensions of the wall frame by referring to the drawing for the assigned structure. What is the overall width and height of the wall frame?

   _____

2. Locate the graph paper provided with this activity. Assume that 1" = 1'. This means that eight 1/8" squares represent one foot. How many inches does each square equal?

   _____

3. Will the wall frame fit within the borderlines of the graph paper, with enough space for dimensions? If not, can you change the scale so the drawing will fit?

   _____

   Try six squares per foot rather than eight. If a change in the scale does not work, check with your instructor.

4. Using light lines, lay out a rectangle equal in size to the width and height of the assigned wall frame. Center the rectangle within the borders of the graph paper.

5. Check the section drawings for the structure to determine the location and thickness of plates. Given the scale you selected, how many squares will equal the thickness of the plates?

   _____

6. Draw light lines in the correct locations to represent the edges of the plates.

7. Assume that 4' × 8' sheathing will be installed and studs are to be located 16" on center. Drawing with light lines, lay out the wall studs.

Copyright by The Goodheart-Willcox Co., Inc.

8. Does the wall you are drawing include an outside corner or intersecting interior wall? If so, you will need to add the studs for these components, as shown here:

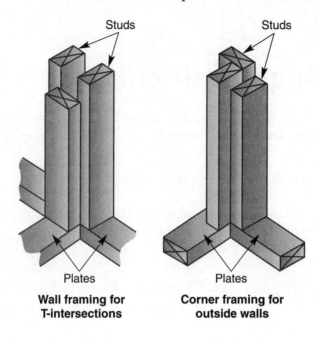

Wall framing for T-intersections

Corner framing for outside walls

9. Is a window or door opening required in the wall frame? If yes, check the structure plans to determine location and size. What is the width and the height? What is the distance from end of wall?

_____

10. Find the distance from one end of the wall frame to the center of the rough opening. Divide the rough opening width by 2 and measure this distance on either side of the rough opening centerline. Check your measurements by measuring the distance between the last two marks. Is this equal to the rough opening dimension? If not, make the necessary corrections.

11. Extend a light lines parallel to the studs to represent the inside edges of the trimmer studs.

12. Add lines outside and parallel to the two lines drawn in Step 11. These lines represent the outside edge of the trimmers. Repeat this step to show the outside edge of the full length studs that must be located outside the trimmer studs.

13. Check the section drawings for the structure to find the width of the header. Measure the header width vertically from the top plate and draw the line representing the bottom edge of the header parallel to the top plate. Note that the header is supported by the trimmer studs and must extend from full length stud to full length stud. How much longer will the header be than the rough opening is wide?

_____

14. If the rough opening is for a window, measure a distance below the header equal to the height of the rough opening and lightly draw the rough sill. The rough sill is made from nominal 2″ thick lumber. This step is not necessary for a door opening.

15. Erase any unnecessary lines and darken the drawing.

16. Add dimensions for stud spacing and for rough opening and overall size.

17. Complete the title block and submit your drawing and this activity for grading.

Name _____

Drawn by:	Date:	Drawing:	Project:

Name _____ Date _____ Class _____

Activity 20-2
Carpentry Activity

Complete Chapter 20 before beginning this activity.

Objectives

Upon completion of this laboratory assignment, you should be able to do the following:

- Correctly and accurately select and cut wall frame components square and within plus or minus 1/16" according to the attached drawing, and using proper construction tools and practices.
- Correctly, neatly, and accurately assemble wall frame components square and within plus or minus 1/16" according to the attached drawing, and using proper construction tools, materials, and practices.
- Lay out and construct exterior and interior wall frames.
- Correctly follow the procedures section of this assignment, submitting your work for inspection and evaluation by the instructor as indicated.
- Carefully disassemble and salvage all components after final evaluation.
- Maintain a clean, orderly work area, properly clean and store tools and machines, and return unused materials to storage.

Tools and Equipment

- One claw hammer
- One framing square or try square
- One folding rule or steel tape measure
- One crosscut saw
- One pencil

Materials and Supplies

- 6d, 10d, and 16d box nails
- One piece of 3/8" plywood, 32 3/4" × 16 3/4", available from instructor
- 2 × 4 lumber

Safety Notes

- Handle lumber, plywood, or OSB carefully. If you get a splinter or a cut, tell your instructor.
- Get help when handling large panels and long boards. Long boards can cause serious injury if they strike another student and can damage property.

Copyright by The Goodheart-Willcox Co., Inc.

- Follow your instructor's directions for sawing, the type of saw to use, and the safety precautions.
- Driving nails can result in smashed fingers. Start the nail with a firm tap, then move your hand away from the nail as the nail is driven.
- Wear eye protection to protect yourself from wood chips and other airborne particles.

Procedural Notes

Read the following procedures, carefully examine Figure A and complete Step 1 before class. Research any unfamiliar terminology and technical information so you are prepared to work when you enter the laboratory.

As you work through the activity and complete each step, make a checkmark in the blank that precedes it. Steps preceded by a circle require your instructor's approval before continuing with the next step. If you do not obtain your instructor's approval, you will not receive full credit for the activity.

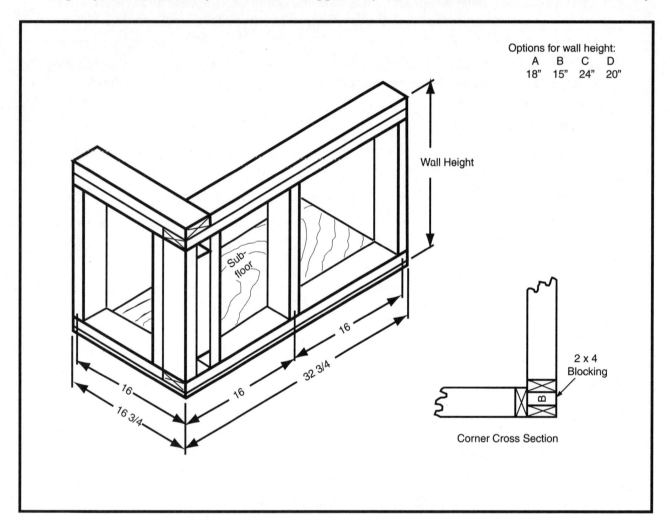

Name _____

Procedure _____

_____ 1. Refer to the drawing to complete the bill of materials. Be sure to account for each component of the frame. All components of the frame are nominal 2 × 4 lumber. Consider the finished dimensions in determining the size of each part of your frame.

Bill of Materials			
Part Name	Quantity	Size	Type of Material
Subfloor (provided)	1	3/8" × 16 3/4" × 32 3/4"	Plywood

_____ 2. Using the bill of materials form, lay out the length dimension of one part on 2 × 4 stock using a rule, try square, and pencil. Have your instructor check your dimensions.

_____ 3. Using the crosscut saw, cut your parts correctly and square.

_____ 4. Repeat Steps 2 and 3 for each part, separately cutting and measuring each part. Do not lay out several parts on one piece of stock and then cut out the parts. If you do this, all subsequent parts will be short.

_____ 5. After all the parts are cut, label each in pencil and ask your instructor to evaluate your work.

_____ 6. Begin assembling the long wall section. Using the plan, lay out the position of the studs on the sole and top plate. Begin measuring from the corner. Check your measurements.

_____ 7. Assemble the corner of the frame using 10d box nails.

_____ 8. Nail the corner and the studs to the long wall sole plate. Nail from the bottom of the plate into the studs and corner of the frame using 16d nails. Make sure that the studs are square to the sole plate.

_____ 9. Nail the top plate to the studs using the same procedures as in Step 8. Make sure the frame is square.

_____ 10. Using the plan, lay out the stud locations on the short wall sole plate and top plate.

_____ 11. Nail the studs to the sole plate and top plate using 16d common nails. Ensure that the studs are square.

_____ 12. Obtain the precut plywood subfloor from your instructor.

_____ 13. Align the sole plate of the long wall along the long edge of the subfloor. Nail the wall frame to the subfloor using 6d nails.

_____ 14. Nail the short wall frame to the subfloor using 6d nails. Be sure the wall frames are square to the subfloor and square to each other.

_____ 15. Nail the side wall frame to the corner using 10d nails. Check that the frame is square.

_____ 16. Nail the double plate on top of the side wall frame. Refer to your plans for the proper location of the double plate. Use 10d nails.

◯ 17. Nail the double plate to the front wall according to the plan. Use 10d nails. Check that the frame complies with the plan specifications and is square. Ask your instructor to evaluate your completed frame model.

◯ 18. Carefully disassemble the wall frame, salvaging as many parts as possible. Neatly stack the salvaged parts in the wood rack.

◯ 19. Complete the student column of Activity 20-3, Carpentry Activity Evaluation. Then ask your instructor to complete the instructor's column.

Name _____ Date _____ Class _____

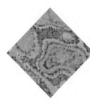

Activity 20-3
Carpentry Activity Evaluation

Read each statement carefully. Decide how many of the possible points you earned and record this number in the student column. Then ask your instructor to evaluate your work.

Evaluation Standards	Maximum Points	Student	Instructor
1. Were all frame components cut accurately (within 1/16"), to dimension and square?	16		
2. Were corners assembled accurately?	5		
3. Was wall frame assembled accurately, according to the plans, and square?	14		
4. Was material properly salvaged? Were all nails removed?	10		
5. Were safety practices followed?	5		
Total Points	50		

Name _____ Date _____ Class _____

Activity 20-4
Observation Form—Miter Saw

Read this observation entirely before working with a learner. Then follow each step as you conduct the practice session for the learner. As you complete each step, make a checkmark in the blank that precedes each step.

_____ 1. Set up the tools and supplies listed below in the work area.

- Wrench for saw blade installation and removal.
- Tape measure
- Try square
- Face shield
- One piece of 2″ × 4″ stock, at least 24″ long
- Bench duster
- A sharp combination blade that fits the miter saw.

_____ 2. Adjust the miter saw so it is in the following condition:

- Unplugged.
- Guard removed.
- Blade removed with the nut and washer placed near it.
- Set to cut at 90° to the fence.

Also dust the table and make sure nothing will interfere with the stock firmly contacting the fence.

_____ 3. Ask the learner to install the blade on the saw. Observe the procedures listed in the table, while the blade is being installed. Record your observations. Refrain from giving any assistance until the assembly is complete or the learner is unable to proceed.

Procedure	Yes	No
A. Teeth of blade pointing in the correct direction.		
B. Washer installed between blade and nut.		
C. Nut tightened firmly using the appropriate tool.		
D. Guard properly positioned and secured.		

Copyright by The Goodheart-Willcox Co., Inc.

_____ 4. If there are any errors, ask the learner to correct the errors. Once all errors have been corrected, proceed to step 5.

_____ 5. Have the learner mark the 2″ × 4″ stock for a right angle cut approximately 2″ from one end of the board. The mark should not be made through a knot or other obstruction.

_____ 6. Direct the learner to cut the board so that the kerf is on the waste side of the line (shorter piece of stock). As the learner is working, observe the following operations and record your evaluation. Stop the operation immediately if you see any safety issues or other problems. Call these issues to the attention of the learner in a way that will not distract or surprise the learner from the work being done.

Procedure	First Try		Second Try	
	Yes	No	Yes	No
A. Loose clothing removed or secured.				
B. Eye protection used without being told.				
C. Stock positioned firmly against the fence without being told.				
D. Left hand at least 3″ from path of the saw blade without being told.				

Do not plug the saw in until A–D are correctly done.

Procedure	First Try		Second Try	
	Yes	No	Yes	No
E. Was cut made on the waste side of the line?				
F. Rate of feed too slow.				
G. Rate of feed too fast.				
H. Saw turned off.				
I. Scrap pushed away with larger piece of stock.				

_____ 7. Review your evaluations recorded in Step 6 with your learner.

_____ 8. If errors were made, be sure your learner understands what they were and how to correct them. Then ask your learner to repeat Steps 5 and 6. Record your evaluations in the appropriate location in the tables.

_____ 9. Ask your learner to sign and date this form.

I have completed this miter saw practice and reviewed the evaluation with my student trainer.

Learner _____ Date _____

Chapter 21 Review
Roof and Ceiling Framing

Name _____ Date _____ Class _____

Study Chapter 21 of the text, then answer the following questions.

_____ 1. ____ framing is made using individual rafters and ceiling joists.
- A. Nonconventional
- B. Engineered
- C. Conventional
- D. Manufactured

_____ 2. The horizontal chord at the bottom of the truss functions as a(n) ____.

Match the type of roof with the correct description.

_____ 3. Roof with one sloping surface   A. Gable
_____ 4. Roof with two sloping surfaces   B. Hip
_____ 5. Roof with one horizontal surface   C. Flat
_____ 6. Roof with four sloping surfaces   D. Shed
                                                                  E. Gambrel

7. Identify each part of the W-truss shown.

A. _____   F. _____
B. _____   G. _____
C. _____   H. _____
D. _____   I. _____
E. _____

_____ 8. To determine the spacing between _____, check the plans and specifications.

_____ 9. True or False? You must know how far roof sheathing will overlap the end wall of a building before you lay out the trusses.

_____ 10. Before the first truss is put in place it is necessary to erect _____.

_____ 11. When installing a truss, the truss must be aligned with marks on the double plates and the truss ends must extend the correct distance beyond the _____ on each side of the building.

12. Define the term *run*.

   _____

   _____

13. Identify each part of the gable roof.

   A. _____   E. _____

   B. _____   F. _____

   C. _____   G. _____

   D. _____

Name _____

Match each of the following roofing terms with the correct description.

_____ 14. A horizontal board that is attached to the tail of each rafter.

_____ 15. A wedge-shaped notch that is cut from the rafter so it seats on top of the double plate.

_____ 16. A piece of wood used to fill a gap, level a surface, or adjust a fit.

_____ 17. The incline of a roof based on a 12" unit.

_____ 18. The horizontal cut for a bird's mouth.

_____ 19. Line length of an overhang.

_____ 20. A horizontal board that forms the peak of the roof.

A. Slope
B. Overhang
C. Ridge board
D. Extension
E. Shim
F. Fascia
G. Seat cut
H. Bird's mouth

Name _____ Date _____ Class _____

Activity 21-1
Rafter Layout

Learning to lay out a common rafter will help you understand roof framing procedures and make it possible for you to build a gable roof. Two terms are used in the following descriptions that may be new to you. *Plumb cut* is used to identify lines or cuts that will be vertical when the rafter is installed. *Bird's mouth* is the V-shaped notch that fits over the double plate. The horizontal cut of the bird's mouth makes it possible to toenail the rafter to the double plate.

You will need the following materials and tools to complete this activity:

- Tape measure
- Framing square
- 2 × 4 or 3 1/2″ strip of paper, length calculated in Step 3
- Pencil
- Eraser

1. Your instructor will assign you one of the options shown in **Figure A** for a common rafter layout.

Figure A—Common Rafter Layouts			
Option	Run	Slope	Overhang
A	3′ 6″	4 on 12	9″
B	4′ 3″	5 on 12	12″
C	3′ 9″	6 on 12	6″
D	4′ 6″	4 on 12	4″
E	3′ 3″	6 on 12	9″

2. What is the rise to run ratio of your assigned rafter?

_____

3. On the wider leg of the framing square, find the row that lists the "length common rafter per foot run." Follow this row to the inch mark equal to the rise of the rafter you are laying out. For example, if the slope is 5 on 12 (5" for each 1' of run), you will find the length of a common rafter per foot of run in the column below the 5" mark. The value for a 5 on 12 slope is 13". Thirteen inches is the length of the hypotenuse of a right triangle with a 12" base and height of 5". Tables such as these eliminate the need to calculate this value. Since the sides of triangles with equal angles are proportional, the length of the hypotenuse of any 5 on 12 rafter can be calculated simply by multiplying the run, measured in feet, times the length per foot of run. What is the length per foot of run for your rafter?

___

What is the length of your rafter? Show your calculations in a neat, orderly fashion.

4. To estimate the 2 × 4 or paper length needed to lay out the common rafter, add material for the overhang plus 6" to the line length. Show your calculations below.

5. Obtain the appropriate length of 2 × 4 or strip of paper. Position the framing square to lay out the plumb cut near one end of the board or paper strip, **Figure B**.

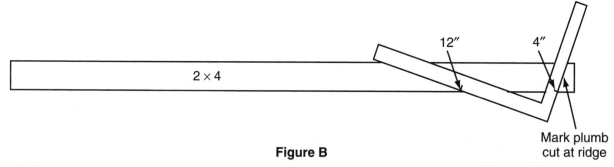

Figure B

6. Measure the line length along one edge and mark the location for the plumb cut for the bird's mouth, **Figure C**.

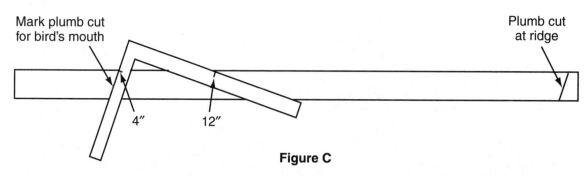

Figure C

Name _____

7. The horizontal cut for the bird's mouth is the same length as the double plate is wide. In this case, use 3 1/2″. To draw this line, position the framing square as shown in **Figure D**.

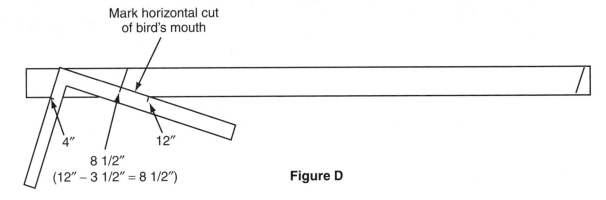

Figure D

8. It may help to hold the rafter so it is sloped as it will be in the roof frame. Is the plumb cut at the ridge vertical and parallel to the plumb cut for the bird's mouth? Is the horizontal cut for the bird's mouth level and 3 1/2″ long? If not, correct the layout before darkening the lines that form the bird's mouth.

9. Add the specified overhang by measuring the overhang distance from the plumb cut of the bird's mouth and lay out a third plumb cut for the overhang end of the rafter, **Figure E**.

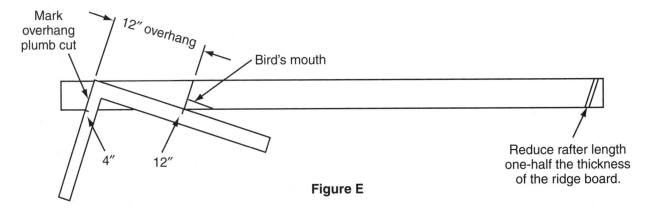

Figure E

10. Shorten the top end of the rafter to allow space for the ridge board. This means that the first plumb cut marked at the top end of the rafter needs to be moved one-half the thickness of the ridge board. Assume that the ridge board will be nominal 1″ lumber. Draw this line and label it *plumb cut*.

11. If you did your layout on a 2 × 4, follow your instructor's directions for having your work graded. If you used a strip of paper for the layout, roll the paper, and secure the roll with a rubber band. Write your name on the outside of the roll and submit the roll along with this activity sheet.

Name _____ Date _____ Class _____

Activity 21-2
Observation Form—Radial Arm Saw

Read this observation form entirely before working with your learner. Then follow each step as you conduct the practice session for the learner. As you complete each step, make a checkmark in the blank that precedes each step.

_____ 1. Set up the tools and supplies listed below in the work area.
- Wrench for saw blade installation and removal
- Tape measure
- Try square
- Face shield
- One 2" × 4", at least 24" in length
- Bench duster
- One sharp crosscut or combination blade that fits the radial arm saw

_____ 2. Adjust the radial arm saw so it is in the following condition:
- Unplugged.
- Guard removed.
- Blade removed with the nut and washer placed near it.
- The over arm set at 90° to the fence.

In addition, dust the tabletop and check the condition of the saw fence. It must be in usable condition.

_____ 3. Ask the learner to install the blade on the saw. Observe the procedures listed in the table, while the blade is being installed. Record your observations. Refrain from giving any assistance until the assembly is complete or the learner is unable to proceed.

Procedure	Yes	No
A. Teeth of blade pointing in the correct direction.		
B. Washer installed between blade and nut.		
C. Nut tightened firmly using the appropriate tool.		
D. Guard properly positioned and secured.		

Copyright by The Goodheart-Willcox Co., Inc.

_____ 4. If there are any errors, ask the learner to correct the errors.

_____ 5. Have the learner mark the 2″ × 4″ stock for a right angle cut approximately 2″ from one end of the board. The mark should not be made through a knot or other obstruction.

_____ 6. Direct the learner to cut the board so that the kerf is on the waste side of the line (shorter piece of stock). As the learner is working, observe the following operations and record your evaluation. Stop the operation immediately if you see any safety issues or other problems. Call these issues to the attention of the learner in a way that will not distract or surprise the learner from the work being done.

Procedure	First Try		Second Try	
	Yes	No	Yes	No
A. Loose clothing removed or secured.				
B. Eye protection used without being told.				
C. Saw completely behind fence before starting, without being told.				
D. Stock positioned firmly against the fence without being told.				
E. Left hand at least 3″ from path of the saw blade without being told.				

Do not plug the saw in until A–D are correctly done.

Procedure	First Try		Second Try	
	Yes	No	Yes	No
F. Was cut made on the waste side of the line?				
G. Rate of feed too slow.				
H. Rate of feed too fast.				
I. Saw returned behind fence.				
J. Saw turned off.				
K. Scrap pushed away with larger piece of stock.				

_____ 7. Review your evaluations recorded in Step 6 with your learner.

_____ 8. If errors were made, be sure your learner understands what they were and how to correct them. Then ask your learner to repeat Steps 5 and 6. Record your evaluations in the appropriate location in the tables.

_____ 9. Ask your learner to sign and date this form.

I have completed this radial arm saw practice and reviewed the evaluation with my student trainer.

Learner _____ Date _____

Chapter 22 Review
Enclosing the Structure

Name _____ Date _____ Class _____

Study Chapter 22 of the text, then answer the following questions.

_____ 1. True or False? Rough-in of plumbing, electrical circuits, and HVAC can begin before the roof is installed.

2. Name five factors that influence roofing material choice.

_____
_____
_____
_____
_____

Match each type of roofing material with the correct description.

_____ 3. Consists of roofing felt, bitumen, and aggregate.

_____ 4. A plant fiber sheet that is soaked in asphalt.

_____ 5. A water-resistant material made of asphalt or tar.

_____ 6. Sheet of special plastic put on a smooth deck.

_____ 7. Sprayed or rolled on and allowed to cure.

_____ 8. Made of asphalt, wood, slate, metal, clay, or concrete.

_____ 9. Flat or curved pieces of material laid in overlapping rows on a roof deck.

A. Bitumen
B. Shingles
C. Liquid roofing
D. Built-up roofing
E. One-ply roofing
F. Roofing felt
G. Flood coat
H. Shingles

_____ 10. A(n) _____ is installed along the edge of the roof deck.

11. What is flashing?

_____
_____
_____

Copyright by The Goodheart-Willcox Co., Inc.

12. Name two safety items that can be used while roofing to prevent falls and to secure footing.

_____

_____

_____ 13. A(n) _____ is the overhang of the roof at the eave or horizontal edge of the roof.
   A. soffit
   B. fascia
   C. cornice
   D. ridge cap

14. Calculate the amount of shingles needed for a gable roof. Each of the two surfaces measures 17′ 9″ × 51′ 3″. Show your work in a neat, orderly fashion to receive full credit.

15. Place the numbers 1–7 in the blanks preceding the steps for installing plywood cornices to indicate the sequence in which they are typically completed.

_____ Cut plywood to width for the soffit.

_____ Nail the 2 × 4 ledger to the studs.

_____ Snap a chalk line the length of the wall.

_____ Nail the soffit in position using corrosion-resistant nails.

_____ Toe nail lookouts to the ledger and face nail them to the sides of the rafter tails.

_____ Nail the frieze to the studs.

_____ Locate the ledger by leveling from the bottom edge of the fascia to the wall. Measure up to allow for the soffit.

Name _____

_____ 16. Structures designed with _____ walls require that these walls be constructed as the building takes shape.
   A. non-load-bearing
   B. curtain
   C. plaster
   D. load-bearing

17. What are masonry walls?

_____

_____

_____

_____ 18. What is the first step in enclosing the exterior walls of a platform-frame building?
   A. Install windows and doors.
   B. Apply house wrap.
   C. Apply masonry or other exterior finish.
   D. None of the above.

19. Identify the parts of the exterior door frame.

   A. _____

   B. _____

   C. _____

20. The first three steps of prehung door installation are described below. Briefly describe, in order, the remaining four steps of this process.

   Step 1. Check the size of the rough opening. Is it the correct size, plumb, and level?

   Step 2. Apply two beads of caulking compound at the bottom of the opening.

   Step 3. Set the pre-hung door in place. Make certain the trim is tight against the wall and the unit is centered horizontally in the opening. Check the threshold for level and shim as necessary.

   Step 4. _____

   Step 5. _____

   Step 6. _____

   Step 7. _____

21. This synthetic building covering made of foam plastic insulation and thin synthetic coating is known as _____.

_____ 22. True or False? Siding is attached directly to sheathing.

23. Place the numbers 1–6 in the blanks preceding the steps in wood siding installation to indicate the sequence in which they are typically completed.

   _____ Caulk and align each succeeding run of siding with the chalk line. Secure the wood siding with corrosion-resistant nails. Drive the nails into each stud about 1/2″ from the bottom edge of each run of siding.

   _____ Caulk joints and nail the first run of siding approximately 1″ from the bottom edge of the siding using rust-resistant nails.

   _____ Determine the exposure of each strip of the siding and prepare a story pole.

   _____ On runs needing more than one strip of siding, cut strips to a length that will permit nailing both pieces to a stud at the point where they butt together.

   _____ Transfer the dimensions from the story pole to each end of a wall section and snap chalk lines indicating the top edge of each strip of siding.

   _____ Install inside and outside corners.

Name _____ Date _____ Class _____

Activity 22-1
Masonry Activity

Complete Chapter 22 before beginning this activity. You will work with a classmate to complete this activity.

Objectives

Upon completion of this laboratory assignment, you should be able to do the following:

- Properly mix a batch of mortar.
- Lay two stretcher courses of block using proper construction practices so the courses are aligned to grade, plumb, and are level within 1/16".
- Properly tool mortar joints.
- Lay two stretcher courses of single tier brick, meeting the same specification as those for the stretcher concrete blocks.
- Maintain a clean, orderly work area, properly clean and store tools, machines, and equipment, return unused materials to storage.
- Carefully salvage brick and block upon completion of the activity.
- Cooperate with classmates.
- Correctly follow the procedures section of this assignment, submitting your work for inspection and evaluation by the instructor as indicated.

Tools and Equipment

- One plywood floor cover with story poles
- One mason's line
- One mason's trowel
- One jointer
- One level
- One rule
- One mixing box (mortar box)
- One hoe
- One shovel
- One water bucket
- One broom
- One mortar board

Copyright by The Goodheart-Willcox Co., Inc.

Materials and Supplies

- Sand
- Water supply
- 3/4″ × 4′ × 6′ plywood for mounting story poles
- 6mm × 6′ × 6′ plastic sheet to cover plywood
- Five stretcher concrete blocks, 8″ × 8″ × 16″
- Two concrete blocks, 8″ × 8″ × 8″
- Three solid blocks, 4″ × 8″ × 16″
- Eleven standard bricks

Safety Notes

- Carry and handle concrete blocks carefully. They are heavy and can cause injury if dropped or broken.
- Wear safety glasses at all times. Masonry cement, lime, sand, dust, and mortar chips can cause serious injuries to your eyes.
- Avoid skin exposure to masonry cement, lime, or mortar. Brief exposure can cause dry skin and extended exposure can cause chemical burns. Wash exposed skin with soap and water immediately.

Procedural Notes

Read the procedures and carefully examine all diagrams before class. **Figure A** shows the wall you will be building. Research any unfamiliar terminology and technical information so you are prepared to go to work in the laboratory.

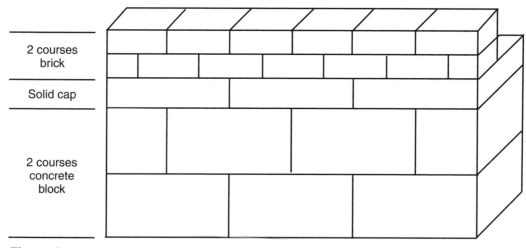

Figure A   Wall to be constructed

Name _____

Procedure

As you work through the activity and complete each step, make a checkmark in the blank that precedes it. Steps preceded by a circle require your instructor's approval before continuing with the next step. If you do not obtain your instructor's approval, you will not receive full credit for the activity.

_____ 1. Position the 3/4" × 4' × 6' plywood sheet in the area designated by your instructor. Cover the plywood with 6mm × 6' × 6' plastic sheet and mount the story poles, **Figure B**.

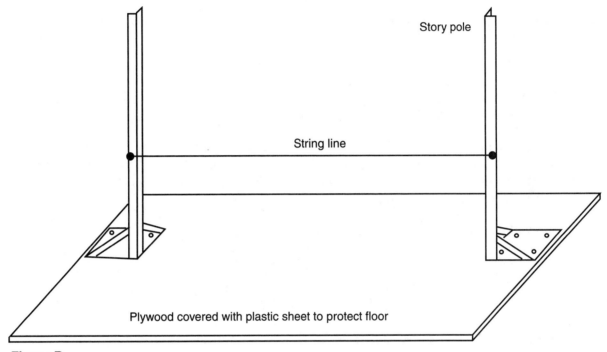

Figure B

_____ 2. Bring all tools, equipment, and materials to the work area.

_____ 3. Mix the mortar in the following proportion:
- Three pints, masonry cement
- Four pints, lime
- Two gallons, sand

This mixture will make a plastic, low-strength, Type 0 mortar. The lime further weakens the mortar, which will allow you to take down the wall without breaking the bricks and blocks.

○ 4. Use the mortar box to dry mix the mortar as demonstrated. Add water until the desired consistency is obtained.

_____ 5. Secure the mason's line to the story poles at the height of the top of the first course of blocks. The line should be level and allow for the mortar bed and floor unevenness.

_____ 6. Spread a full mortar bed for the first course of block and furrow it with a trowel.

Copyright by The Goodheart-Willcox Co., Inc.

_____ 7. Carefully lay the corner block (end block), making sure it is positioned correctly and accurately. It will guide the rest of the course. Be sure this and all other block is laid face shell up. Face shell is the surface with largest area of concrete.

_____ 8. Place two blocks on end and butter the top end of each for a vertical joint.

_____ 9. Lower these two blocks successively into position one after the other. Push them downward into the mortar bed to produce well-filled vertical joints. Clean off excess mortar with the trowel.

_____ 10. Check the alignment and grade using the mason's level. Position the blocks with light taps of the trowel. Use the mason's level to check plumb.

_____ 11. Move the mason's line up for the second course. Ensure that the height and level are correct.

_____ 12. Apply a face shell bedding to the top of the first course of block.

_____ 13. Begin the second course with the 8" × 8" × 8" block flush on top of either end of the first course. As in Step 10, check for grade, alignment, and plumb.

_____ 14. Repeat Steps 8–10, placing two full size blocks and ending with another 8" × 8" × 8" block. You will want the wall to have a stretcher bond (pattern). The top of the second course should be level to within 1/16" of the mason's line or story pole mark. The face should be in line with the first course and plumb.

_____ 15. Move the mason's line up the story pole for the next course.

_____ 16. Apply mortar to the top of the second course of block.

_____ 17. Carefully lay a solid concrete block (4" × 8" × 16") flush on top of either end of the second course. Check plumb, alignment, and grade as in Step 10.

_____ 18. Following the procedures in Steps 8–10, lay the remaining solid concrete blocks.

_____ 19. Check your work as explained in Steps 10 and 14.

_____ 20. Move the mason's line up for the first course of brick.

_____ 21. Place a thick one inch bed of mortar, the width of the bricks, flush with the front edge of the block wall. Make a shallow furrow along the center of the bed. The bed should be about four brick lengths long.

_____ 22. Place the first brick at either end of the wall and tap it into the mortar bed. Check as outlined in Steps 10 and 14.

_____ 23. Butter the head of the succeeding brick and lay it in the mortar bed, pushing it against the first brick so that excess mortar squeezes from the head joint. Clean excess mortar with the trowel.

_____ 24. Repeat Step 21 and Step 23 until the first course of brick is laid. Check your work as described in Steps 10 and 14. Your mortar joints for the brick should be about 3/8".

_____ 25. Repeat Steps 21–24 for the second course of brick, except place the first brick of the second course inward one-half length from the end of the first course of brick, in a stretcher bond.

_____ 26. Tool all joints with a concave jointer.

Name _____

_____ 27. Clean your site, tools, and equipment, place unused materials in the proper storage space, and clean out the mortar box.

◯ 28. Complete the student column of Activity 22-2, Masonry Activity Evaluation. Then ask your instructor to complete the instructor's column.

_____ 29. Carefully dismantle your wall. Wearing safety glasses or goggles, lightly tap the mortar joints with a brick chisel and hammer. Clean all bricks and blocks of mortar and place them in the proper storage area. Ask your instructor if the story poles should be removed and stored. Clean the site.

◯ 30. Ask your instructor to inspect your cleanup.

Name _____ Date _____ Class _____

Activity 22-2
Masonry Activity Evaluation

Read each statement carefully. Decide how many of the possible points you earned and record this number in the student column. Then ask your instructor to evaluate your work.

Evaluation Standards	Maximum Points	Student	Instructor
1. Was the batch of mortar properly mixed?	5		
2. Were the two stretcher courses of block aligned, plumb, and within 1/16″ of level?	13		
3. Were the two courses of brick laid to meet the same specification as in Number 2 above?	13		
4. Were all mortar joints properly tooled?	4		
5. Was work area properly cleaned, with tools and extra materials returned to storage at the end of each work period?	4		
6. Were brick and block carefully salvaged?	4		
7. Were the instructor's initials obtained when required?	3		
8. Were safe work practices followed during this activity?	4		
Total Points	50		

Copyright by The Goodheart-Willcox Co., Inc.

Chapter 23 Review
Plumbing Systems

Name _____ Date _____ Class _____

Study Chapter 23 of the text, then answer the following questions.

1. Name nine types of piping systems that may be installed inside buildings.

   _____
   _____
   _____
   _____
   _____
   _____
   _____
   _____
   _____

_____ 2. Underground pipes that branch off trunk lines to bring water near individual buildings is called a(n) _____.

3. Identify the parts of this municipal water supply hookup.

   A. _____   D. _____

   B. _____   E. _____

   C. _____   F. _____

_____ 4. True or False? Valves are installed to isolate portions of the piping so that repairs can be made without turning off all water to a building.

_____ 5. A drain-waste-vent system carries sewage to _____.

   A. sanitary sewers
   B. storm sewers
   C. nearby bodies of water
   D. None of the above.

6. Sanitary sewers take waste to a(n) _____ plant.

_____

_____ 7. In a DWV system, _____ remain filled with water, preventing sewer gas from escaping into the structure.

   A. vents
   B. floor drains
   C. traps
   D. cleanouts

Name _____

8. What is the purpose of placing horizontal wastewater pipes on a downward slope of 1/4″ for each foot of length?

_____

_____

Match each term with the appropriate description.

_____ 9. Used to treat sewage in areas where sanitary sewers are not available.

_____ 10. Disperses the outflow from a septic tank.

_____ 11. A firefighting system that consists of a piping system connected to a reserve source of water and fire hoses.

_____ 12. Installed in facilities that use air-powered tools.

_____ 13. A container installed in a basement floor to collect groundwater from foundation drains.

A. Standpipe system
B. Septic system
C. Toxic waste
D. Sump
E. Leach field
F. Sprinkler system
G. Compressed air piping system

_____ 14. The design of residential and light commercial plumbing systems is dictated by _____.
   A. architects
   B. plumbers
   C. engineers
   D. plumbing code

15. Name the four grades of copper pipe typically used in water supply and DWV piping.

_____

_____

_____

_____

16. Identify the following copper pressure fittings.

   A. _____

   B. _____

   C. _____

   D. _____

   E. _____

   F. _____

_____ 17. The degree of material hardness is known as _____.

_____ 18. During the first-rough stage of plumbing installation, _____ are installed.
   A. sinks and lavatories
   B. tubs and shower bases
   C. water and sewer mains
   D. kitchen appliances

_____ 19. True or False? Chlorinated polyvinyl chloride pipe is required for cold water.

20. Place the numbers 1–3 in the blanks preceding the following steps in the installation of plastic or copper pipe and their fittings to indicate the sequence in which they are typically completed.

_____ Add a fitting allowance for each pipe end to the face-to-face dimension.

_____ Measure the distance from the face of one fitting to the face of the other.

_____ Locate the pipe size in a chart to determine the fitting allowance.

_____ 21. True or False? The first-rough installation is inspected and tested before the water supply and building sewer pipes are covered.

Name _____

_____ 22. The inspection of the second rough phase of plumbing installation includes checking that _____.
A. vents are correctly installed
B. pipes are properly supported
C. cleanouts are installed as required by code.
D. All of the above.

23. Inspection of work completed during the finish stage of plumbing installation is done as a part of the _____ of the building.

_____

Name _____ **Date** _____ **Class** _____

Activity 23-1
Green Plumbing Products

This assignment focuses on plumbing products that both reduce water consumption and take advantage of renewable energy to heat water. With your instructor's help, select one of the following topics about green plumbing products. Prepare a report, presentation, or display that describes the product and focuses on its strengths and weaknesses compared to materials traditionally used for the same purpose in residential construction.

Possible topics:

- High efficiency toilets.
- Shower heads and faucets that use less water.
- Gray water systems.
- Instantaneous water heaters.
- Solar energy for heating water.
- Other topic approved by your instructor.

Your instructor will provide specific directions for your report regarding things such as length, presentation time, and display size.

Copyright by The Goodheart-Willcox Co., Inc.

Name _____ Date _____ Class _____

Activity 23-2
Plumbing Activity

Complete Chapter 23 before beginning this activity.

Objectives

Upon completion of this laboratory assignment, you should be able to do the following:

- Select, cut and fit PVC and copper pipe and fittings within plus or minus 1/16" according to the attached drawings, using the proper tools and practices.
- Correctly follow the procedures section of this assignment, submitting your work for inspection and evaluation by the instructor as indicated.
- Carefully disassemble and salvage all components after final evaluation.

Tools and Equipment

- Rule
- Acid swab
- Tubing cutter
- Plastic pipe cutter
- Spark lighter
- Propane torch
- Rag

Materials and Supplies

- 120 grit or finer abrasive paper
- Flux
- 50-50 solder
- PVC cleaner or primer
- PVC adhesive

Safety Notes

- Handle abrasive paper with care. It can cut skin and drive slivers of copper into your hands. Remove burrs with a reamer.
- When using a propane torch, point the torch tip away from yourself. Make sure none of your classmates is in the pathway of the torch when it is lit.
- Set a lit torch down carefully. The torch is top heavy and can be easily tipped over.
- After heating copper pipe and fittings, they remain hot and can cause burns if touched. Cool the parts in water or place them on a metal or ceramic surface to cool.
- Soldering flux can harm skin. Use an acid swab to apply flux and wipe the completed joint with a rag to remove any residual flux on the pipe and fitting.

Copyright by The Goodheart-Willcox Co., Inc.

Procedural Notes

Read the following procedures, carefully examine the drawings, and complete Steps 1 and 2 before class. Research any unfamiliar terminology and technical information so you are prepared to work when you enter the laboratory.

As you work through the activity, read each step again carefully. As you complete each step, make a checkmark in the blank that precedes each step. Steps preceded by a circle require your instructor's approval before continuing with the next step. If you do not obtain your instructor's approval, you will not receive full credit for the activity.

_____ 1. Your instructor will tell you which dimension option you should follow to complete this laboratory assignment. Write the option you are assigned here.

_____

○ 2. Refer to each drawing to complete the bills of materials. Account for each component separately and for fitting allowances when determining the length of pipe. Have your instructor check your bill of material and initial this step.

Dimensions	Options		
	1	2	3
A	12"	10"	8"
B	6"	4"	4"
C	12"	10"	8"
D	4"	6"	6"

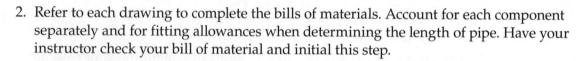

1/2" PVC pipe and fittings

PVC Plumbing Activity—Bill of Materials		
Quantity	Size	Description

Name _____

Dimensions	Options		
	1	2	3
A	12"	10"	8"
B	3"	4"	3"
C	3"	3"	3"
D	2"	3"	2"

1/2" Copper Pipe and Fittings

Copper Plumbing Activity—Bill of Materials		
Quantity	Size	Description

_____ 3. Proceed to either the PVC or copper plumbing activity.

PVC Plumbing Activity _____

_____ 1. Using the PVC bill of materials, lay out each length of PVC pipe.

_____ 2. Using the PVC pipe cutter, cut each piece of pipe to length.

○ 3. Ream each end of each length of pipe.

_____ 4. Clean the inside of each fitting socket and the ends of each pipe using PVC cleaner or primer.

_____ 5. Apply PVC adhesive to one end of a length of pipe and to the inside of the appropriate fitting. Assemble the pipe and fitting by turning the pipe approximately one-third as you insert it fully into the fitting. Repeat this procedure for each joint, making sure that the finished assembly matches the drawing.

○ 6. Rate your work in the student column of the Plumbing Activity Evaluation, Activity 23-3. Submit the finished assembly along with evaluation sheet to your instructor for review.

_____ 7. Proceed to the copper exercise if you have not already completed it.

Copper Plumbing Activity

_____ 1. Using the copper bill of materials, lay out each length of copper pipe.

_____ 2. Using a tubing cutter, cut each piece of pipe to length.

○ 3. Ream the end of each length of pipe. Check each pipe for burrs. Ask your instructor to inspect your work.

_____ 4. Use fine abrasive paper to clean the end of each pipe and the inside of the fittings.

_____ 5. Apply flux to the end of one pipe and the inside of the appropriate fitting. Assemble the joint.

_____ 6. Heat and apply solder to the joint. Carefully wipe the joint with a damp rag to remove excess solder. Cool the joint. Repeat this procedure for each joint, making certain that the assembly matches the drawing.

○ 7. Rate your work in the student column of the Plumbing Activity Evaluation, Activity 23-3. Submit the finished assembly along with evaluation sheet to your instructor for review.

_____ 8. Proceed to the PVC activity if you have not already completed it.

Salvage

Your instructor will return your PVC and copper assemblies to you after they have been evaluated. You will need to salvage the materials by following the procedure given here:

_____ 1. Cut the fittings off the PVC assembly. Save all parts of the assembly.

_____ 2. Disassemble the copper pipe and fittings and remove as much of the solder as possible.

○ 3. Rate your work in the salvage section of the Plumbing Activity Evaluation form, Activity 23-3. Ask your instructor to check your salvage work and to complete the evaluation in Activity 23-3.

_____ 4. Place the PVC fittings in the proper salvage bin. Return all other materials to the proper storage location.

Name _____ Date _____ Class _____

Activity 23-3
Plumbing Activity Evaluation

Read each statement carefully. Decide how many of the possible points you earned and record this number in the student column. Then ask your instructor to evaluate your work.

Evaluation Standards	PVC Pipe and Fittings			Copper Pipe and Fittings		
	Maximum Points	Student	Instructor	Maximum Points	Student	Instructor
1. Measurements within plus or minus 1/16".	5			5		
2. Fittings located to produce correct angle relationships.	4			4		
3. Pipe cut square.	2			2		
4. Pipe correctly reamed.	3			3		
5. Joints correctly made.	5			5		
Salvage						
6. Parts disassembled and cleaned.	2			2		
7. Materials properly stored.	2			2		
8. Were safety practices followed?	2			2		
Total Points	25			25		

Copyright by The Goodheart-Willcox Co., Inc.

Chapter 24 Review
Heating, Ventilating, and Air-Conditioning (HVAC) Systems

Name _____ Date _____ Class _____

Study Chapter 24 of the text, then answer the following questions.

Match each HVAC components with the correct description.

_____ 1. Controls air circulation.

_____ 2. Removes dust and other contaminants.

_____ 3. Adds moisture to the air.

_____ 4. Controls temperature.

_____ 5. Removes moisture from the air.

A. Heating and cooling unit
B. Humidifier
C. Dehumidifier
D. Air exchangers and ventilator
E. Filters and electronic air cleaner
F. Regulator

6. Name three energy sources commonly used to control indoor temperatures.

_____

_____

_____

_____ 7. Heat produced by the sun's rays is called _____ energy.

_____ 8. True or False? Heat pumps burn fuel to create heat.

Match the type of heating unit with the correct description.

_____ 9. Includes a blower, burner, heat exchanger or electric heating element, and controls.

_____ 10. Uses a pipe buried eight to ten feet below the ground surface.

_____ 11. Uses refrigerant to transfer heat.

_____ 12. Water passes through pipes or tubes to convectors located in each room.

_____ 13. Has embedded resistance wires placed in walls, floors, or ceilings.

_____ 14. Uses fans or pumps to move the heated air or water.

_____ 15. Heat is collected within masses of materials

_____ 16. A(n) _____ is a mechanical device that removes heat from the air in a structure.
   A. solar collector
   B. heat pump
   C. air conditioner
   D. furnace

A. Active solar collector
B. Solar module
C. Passive solar collector
D. Hydronic heating system
E. Radiant heat system
F. Electric furnaces
G. Furnace
H. Geothermal system
I. Heat pump

17. Identify the air-conditioning components.
   A. _____
   B. _____
   C. _____
   D. _____
   E. _____
   F. _____
   G. _____

Name _____

_____ 18. The _____ absorbed or given off when refrigerant changes state is the key to operation of air conditioners and heat pumps.

_____ 19. True or False? The latent heat required to change a liquid to a gas changes the temperature of gas compared to the liquid.

_____ 20. The amount of heat required to cause a refrigerant to change state is known as _____ heat.
   A. real
   B. latent
   C. active
   D. potential

_____ 21. A typical forced-air heating and cooling system has _____ systems.

_____ 22. The supply-air duct system includes the _____.
   A. plenum
   B. extended plenum
   C. branch pipes
   D. All of the above.

23. How is water heated in hydronic and steam heating systems?

_____

24. Name the type of heating system that uses resistance wires.

_____
_____

_____ 25. True or False? Hot air sinks and cold air rises.

_____ 26. A device that regulates the setpoint temperature in a building is called a(n) _____.
   A. zone control system
   B. dehumidifier
   C. humidifier
   D. thermostat

27. The amount of water vapor in the air at a given temperature is called _____.

_____

28. Place the numbers 1–4 in the blanks preceding the steps in HVAC installation in the order they are typically completed.

_____ Make connections to appropriate electrical circuits.

_____ Place the heat exchanger in the proper location outside the building.

_____ Put the furnace in place.

_____ Install a drain to remove condensation that collect on the condenser.

Name _____ Date _____ Class _____

Activity 24-1
Green HVAC Products

This assignment focuses on HVAC products that reduce energy consumption and take advantage of renewable energy for heating and cooling buildings. With your instructor's help, select one of the following tropics about green HVAC products. Prepare a report, presentation, or display that describes the product and focuses on its strengths and weaknesses compared to products traditionally used for the same purpose in residential construction.

Possible topics:

- High efficiency gas furnaces.
- Heat pumps.
- Geothermal systems.
- Passive solar heating systems.
- Active solar heating systems.
- Other topic approved by your instructor.

Your instructor will provide specific directions for your report regarding variables such as length, presentation time, or display size.

Chapter 25 Review
Electrical Power Systems

Name _____ Date _____ Class _____

Study Chapter 25 of the text, then answer the following questions.

1. What is a power system?

   _____
   _____
   _____

2. Electricity moves in a set route from generation to buildings. Number the specific parts of the distribution system in the order in which electricity flows through them.

   _____ Distribution station

   _____ Power generating plant

   _____ Service transformer

   _____ Service cable

   _____ 3. True or False? The service transformer increases the voltage of the electricity coming from the distribution station.

   _____ 4. Distribution lines begin at a(n) _____ and extend to a(n) _____.
   A. distribution station, a service transformer
   B. distribution station, a service mast
   C. power generating plant, service transformer
   D. power generating plant, distribution station

5. Identify the parts of the service entrance.

    A. _____
    B. _____
    C. _____
    D. _____
    E. _____

_____ 6. True or False? The main breaker shuts off power to more than one building.

_____ 7. Which of the following is *not* a part of a branch circuit?

    A. electrical receptacles
    B. appliances
    C. wiring
    D. lights

Match each component with the correct description.

_____ 8. Two or more insulated wires bundled together.

_____ 9. Protects electrical wiring from electrical overload.

_____ 10. Material that allows free movement of electricity.

_____ 11. Interrupts the circuit in the fraction of a second.

_____ 12. Tubes or pipes used to encase and protect electrical wire.

_____ 13. Impedes the flow of electrical current.

A. Fixed-appliance circuit
B. Circuit breaker
C. Handy box
D. Conduit
E. Conductor
F. Cable
G. Insulation
H. Ground-fault circuit interrupter

Name _____

_____ 14. True or False? Color-coded insulation indicates the function of the conductor.

15. Identify the following electrical components.

A. _____

B. _____

C. _____

D. _____

E. _____

F. _____

_____ 16. Which of the following statement regarding receptacles is *not* true?
  A. The duplex receptacle is the most commonly used.
  B. Locking receptacles are useful in places where vibrations can loosen a connection.
  C. They are also known as *outlets*.
  D. They open and close a circuit.

Name _____

17. Name four types of switches.

   _____
   _____
   _____
   _____

_____ 18. The power company is responsible for installing the _____.
   A. meter base
   B. service mast
   C. wire from the pole to the meter
   D. All of the above.

_____ 19. An electrical subcontractor installs the _____.
   A. meter base
   B. service mast
   C. service panel
   D. All of the above.

_____ 20. The electrical code in most communities is based on the _____.
   A. National Electrical Code
   B. International Electrical Code
   C. State Electrical Code
   D. US Electrical Code

Name _____ Date _____ Class _____

Activity 25-1
Electrical Activity

Complete Chapter 25 before beginning this activity.

Objectives

Upon completion of this laboratory assignment, you should be able to do the following:
- Prepare a schematic drawing for an electrical circuit.
- Properly install electrical boxes, cable, receptacles, switches, and a light socket to produce a functioning electrical circuit.
- Work safely with electricity.

Tools and Equipment

- One frame from carpentry activity
- Hammer
- Phillips screwdriver
- Wire cutter
- Wire stripper
- Needle nose pliers
- 1/2" electric drill
- Electrician's auger or spade bit
- Electrical receptacle tester
- Volt-ohm meter

Materials and Supplies

- One 2-gang outlet box
- Two outlet boxes
- Twelve fasteners for attaching boxes
- One octagonal box
- One single-pole switch
- Two three-way switches
- One light socket
- Two duplex receptacles
- One male plug
- Seven cable clamps
- Five 14-2 with ground cables
- One 14-3 with ground cables
- Twelve cable staples
- One lightbulb

Copyright by The Goodheart-Willcox Co., Inc.

254 Construction and Building Technology Tech Lab Workbook

Safety Notes

Electrical shocks can be deadly. They can cause burns and create reactions that cause other problems, such as falls and cuts. Do not connect the circuit to a power source without your instructor's approval. Disconnect the circuit before making any changes or disassembly.

When working on a circuit in your home or on the job, turn off the circuit at the circuit breaker to accomplish the same thing. Ensure that the correct circuit was turned off before attempting to work on the circuit.

Procedural Notes

Read the following procedures, carefully examine all drawings, and complete Steps 1 and 2 before class. Research any unfamiliar terminology and technical information so you are prepared to work when you enter the laboratory.

As you work through the activity, read each step again carefully. As you complete each step, make a check mark in the blank that precedes each step. Steps preceded by a circle require your instructor's approval before continuing with the next step. If you do not obtain your instructor's approval, you will not receive full credit for the activity.

Procedure

_____ 1. Study the illustration below. It shows the location of the boxes, light switches, and outlets in the wood frame.

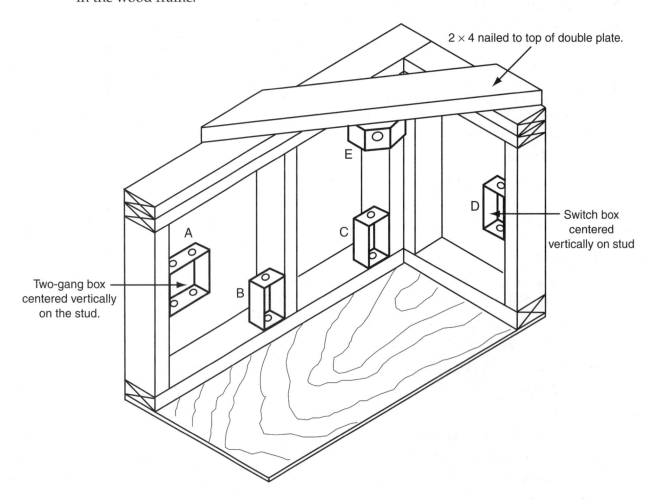

Name _____

The boxes will be wired as follows:

- Box A—Three-way switch controlling ceiling light in Box E.
- Box A—Single-pole switch controlling upper receptacle in Box B.
- Box B—Duplex receptacle. Upper receptacle controlled by single-pole switch in Box A. Lower receptacle not switched.
- Box C—Duplex receptacle not switched.
- Box D—Three-way switch controlling ceiling light in Box E.
- Box E—Ceiling light controlled by three-way switches in Boxes A and D.

2. Complete the schematic drawing below. Ask your instructor to check your drawing before proceeding to the next step. In the drawing, specify the following:

    - The type of cable run between each of the boxes.
    - How the black, white, and red conductors will be connected to the duplex receptacles, switches and light socket.

    Note the following regarding the drawing:

    - Wiring will be done with 14-2/ground and 14-3/ground cables.
    - Draw and label black (B), white (W), red (R), and ground (G) wires parallel to one another outside the boxes.
    - Cable enters boxes through openings.
    - Wires extend to switches and receptacles outside the boxes.
    - Draw lightly in pencil and check your drawing before darkening lines.

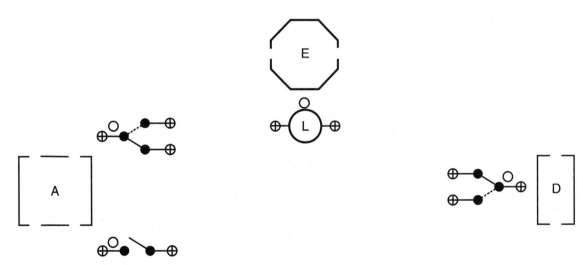

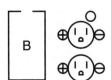

LEGEND:
Connecting screws:
⊕ Brass—black or red wires:
⊖ Chrome—white wires:
○ Green—safety ground wires

_____ 3. Begin by installing the electrical boxes in the frame. Attach a 2 × 4 diagonally across the top of the double plates. This will hold the octagonal box for the light socket.

_____ 4. Drill holes to run cable between the boxes. Drill these holes as near the center of the framing members as possible.

_____ 5. Install cable clamps and cable as required to wire the circuit as shown in the schematic drawing. Secure the cables with staples. Do *not* connect your circuit to an electrical power source until the circuit has been inspected by your instructor (Step 6).

◯ 6. Attach the switches, light socket, and duplex receptacles. Check your work carefully. Ask your instructor to review your work and initial the circle, indicating that you may install a lightbulb and connect your circuit to 120V electrical power as directed by your instructor.

_____ 7. Check the wiring of the duplex receptacles using an electrical receptacle tester. Does the switch control the top socket of the duplex receptacle? Can the light be turned on or off from either of the three-way switches? Use the volt-ohm meter to check the voltage available at each outlet.

_____ 8. Disconnect the power source and correct any problems. Ask your instructor to demonstrate how to use the volt-ohm meter to check wiring when the power is disconnected.

◯ 9. Have your instructor recheck and approve your work before connecting to the power source. If problems still exist, repeat Steps 8 and 9 until everything is working correctly. Always disconnect the power source before working on the circuit.

◯ 10. Complete the student column of the Electrical Activity Evaluation form. Then ask your instructor to complete the instructor's column.

_____ 11. Remove the switches, receptacles, light socket, cables, and boxes and return these items to storage. Disassemble or place the frame in storage as directed by your instructor.

_____ 12. Put away any tools, materials, or supplies and clean the work area.

Name _____ Date _____ Class _____

Activity 25-2
Electrical Activity Evaluation

Read each statement carefully. Decide how many of the possible points you earned and record this number in the student column. Then ask your instructor to evaluate your work.

Evaluation Standards	Maximum Points	Student	Instructor
1. Was the schematic drawing correct?	7		
2. Were the boxes correctly located and secured?	7		
3. Were the cables located and secured properly?	7		
4. Were connections properly made?	7		
5. Did the switches control the lights and outlet as designed?	7		
6. Were safety precautions followed?	5		
7. Was the work area properly maintained?	5		
8. Were materials and supplies properly salvaged and returned to storage?	5		
Total Points	50		

Chapter 26 Review
Communication Systems

Name _____ Date _____ Class _____

Study Chapter 26 of the text, then answer the following questions.

_____ 1. True or False? All communication systems use a wiring network.

Identify each of the following systems as a monitoring system (M) or an exchange system (E).

_____ 2. Surveillance system

_____ 3. Telephones

_____ 4. Computers

_____ 5. Water quality testing system

_____ 6. A(n) _____ system provides two-way voice and data communication.

_____ 7. True or False? The signal from a cell phone goes directly to the local phone company switching equipment.

Match each communication device with the correct description.

_____ 8. Used for voice, and sometimes visual, communication within a structure.

_____ 9. The part of the computer that processes information.

_____ 10. The location in a building where all telephone wiring is connected to the building.

_____ 11. An electronic device that converts a video signal back into an image with sound.

_____ 12. A device that shields conductors carrying electronic signals from interference.

_____ 13. A telephonic device that sends and receives printed images.

_____ 14. A telephone cable that connects a building to the local telephone network.

A. Coaxial cable
B. Fax machine
C. Intercom system
D. Modem
E. Drop
F. Terminal block
G. Monitor
H. Central processing unit
I. Software

Copyright by The Goodheart-Willcox Co., Inc.

15. Write the name of each part of this television system in the corresponding space.

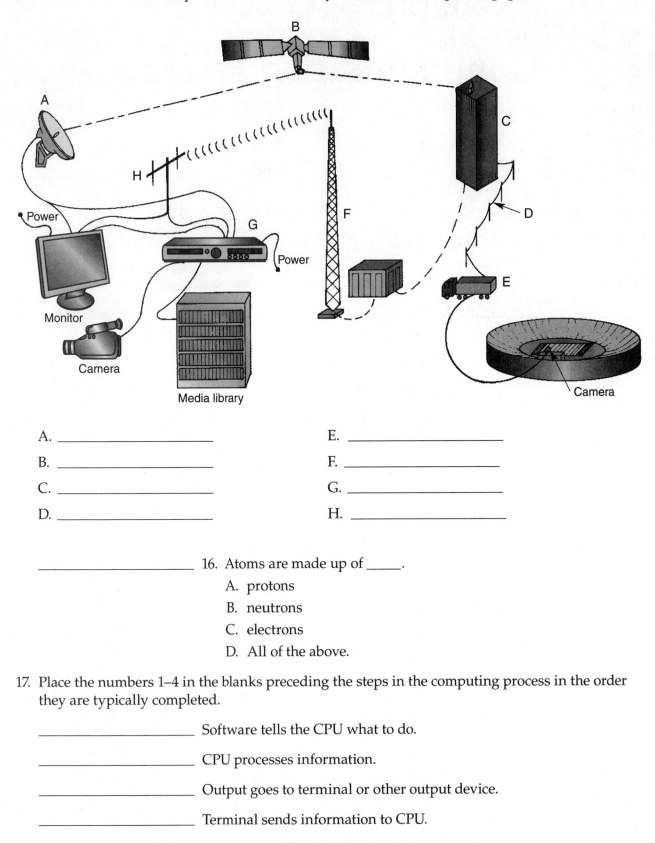

A. _____ E. _____
B. _____ F. _____
C. _____ G. _____
D. _____ H. _____

_____ 16. Atoms are made up of _____.
  A. protons
  B. neutrons
  C. electrons
  D. All of the above.

17. Place the numbers 1–4 in the blanks preceding the steps in the computing process in the order they are typically completed.

_____ Software tells the CPU what to do.

_____ CPU processes information.

_____ Output goes to terminal or other output device.

_____ Terminal sends information to CPU.

Name _____

18. What is the major difference between communication system wiring and electrical power wiring?

   _____
   _____
   _____
   _____

_____ 19. True or False? Each type of communication system wiring is installed separately.

_____ 20. To ensure that the correct types and sizes of wire have been installed, a(n) _____ inspection is done.

Name _____ **Date** _____ **Class** _____

Activity 26-1
Installing Communication Products

Doorbells, intercoms, telephones, security systems, and cable and dish television are all important parts of the communications systems installed in residential structures. Many of these systems are connected to the electrical power system and use transformers to reduce 120-volt electrical current to the lower voltage needed for most communication equipment.

Doorbells

Most doorbells are designed to chime twice when the front doorbell switch is pressed and once when a secondary doorbell is pressed. The bell or chime is placed where it can be easily heard within the home. The switches are placed so they are visible to visitors. The transformer is mounted on an octagonal electrical box and is connected to 120-volt current. The transformer should be located where it is easily accessible for service and replacement. Two-wire cable connects the transformer, push-button switches, and bell or chime. The switches are connected and secured, and the bell or chime is wired and mounted to the wall.

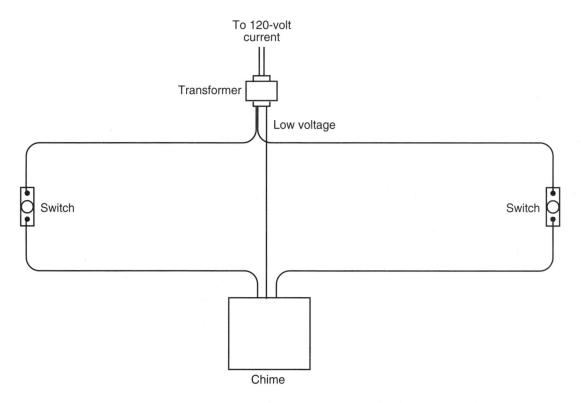

Intercoms, security systems, and satellite and cable television also require 120-volt electrical current. Many of these devices have battery backup to keep them operating even when electrical service is interrupted.

Copyright by The Goodheart-Willcox Co., Inc.

Intercoms

Intercom systems are available as either an independent system or combined with a telephone system that also functions as an intercom. Independent systems are connected to 120-volt electrical current that is reduced to power the intercom system.

Installing an intercom system involves:

- Deciding on the location of the master station and each of the intercom speakers.
- Attaching specially designed brackets to the wall framing at the desired locations.
- Drilling holes in wall framing for cables that will connect the master station and intercom speakers.
- Installing the cables to connect the components.
- Extending a 120-volt cable to the base station to provide electrical current.
- Wiring and mounting the master station and intercom speakers in the brackets or boxes.
- Testing the system.

Telephone Systems

Telephone companies provide access to outside wiring for telephones, but the telephone systems within a building are the responsibility of the building owner.

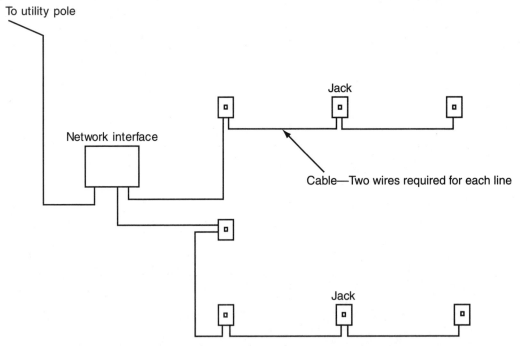

Cat 5 cable containing two pair of twisted wires is typically installed to allow a second telephone line to be connected, without rewiring the jacks.

Installing the interior wiring for a telephone system includes:

- Choosing the locations of boxes for telephone jacks. Wall mounted phone jacks are positioned four feet from the floor. Tabletop phone jacks are mounted 12"–16" from the floor.
- Attaching boxes to the wall framing.
- Drilling holes through wall framing to allow for cable installation.

Name _____

- Installing the telephone cable that connects the outlets to the terminal block. The cable extends at least 6″ outside of each box and is stapled to the wall framing.
- Connecting to the power source provided by the telephone company.
- Wiring jacks and mounting them into the boxes. Attaching cover plates to jacks.
- Testing the system.

Cordless telephones have cordless handsets that function within about 100 feet or so of a base station. The base station is connected to the telephone cable and to a 120-volt electrical outlet. The electricity operates a low-power transmitter and receiver, a battery charger for one handset, and, in some cases, an answering machine. Additional handset bases can be placed anywhere within the range of the master station. They also require an electrical outlet to operate the battery charger for the batteries in the hand set. Dial-up modems and fax machines can also be attached to telephone wiring for Internet access.

Security Systems

Security systems can include sensors, intercoms, and closed circuit television devices. Types of sensors include window attachments that detect breaking glass, switches that detect open doors, and sensors that detect motion, heat, smoke, and water. Closed-circuit television and intercom devices allow residents to see and talk with people at an exterior door. Closed-circuit televisions equipped with recording equipment remotely monitor areas around and inside buildings.

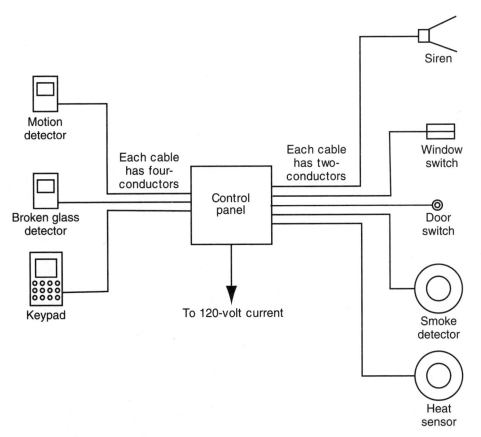

Size and complexity of control panel determines the type and number of detectors and key pads that can be connected to the security system

Copyright by The Goodheart-Willcox Co., Inc.

Once activated, sensors set off alarms within the building and some automatically alert a monitoring service, if the alarm is not disarmed in a set time. The monitoring service contacts police, fire and rescue, or other personnel, who then respond to the emergency.

Security systems are wired in zones of similar types of detectors. This allows automated messages to be sent to the monitoring service indicating the type of emergency that has been detected. Zoned wiring also allows users to deactivate selected zones while keeping other zones active. For example, interior motion detectors can be deactivated when residents are in the house without deactivating other zones.

Cable and Dish Television Service

Cable and dish television service enters a building much like telephone service. The entry cable is connected to a terminal block. Cables run from the block to outlets throughout the building. Cable television may also provide telephone and Internet connection through a single cable. Boxes are connected to the television outlets to control the signal delivered to the television set. The channel selection boxes may also include a recording device. Dish systems require installation of a dish antenna that is focused on an orbiting satellite.

Specific information about the installation of residential communication systems is provided with the products purchased and may also be available on the Internet.

Name _____

Construction and Building Technology

Contractors's License

This is to certify that _____
has satisfactorily passed the performance and written test requirements to prove competence as a:

○ General Contractor

○ Carpentry Subcontractor

○ Concrete Subcontractor

○ Masonry Subcontractor

○ Plumbing Subcontractor

○ Electrical Subcontractor

○ Salvage Subcontractor

This license is valid from the date of certification until the end of salvage.

As a licensed contractor in this class, the holder of this certificate is entitled to bid competitively against other contractors for work within the classification(s) for which the license is granted.

The contractor's responsibilities include planning, organizing, directing and controlling all work contracted for and accounting for all income and expenditures occuring from this work.

Registration as a duly certified contractor was completed on the _____ day of _____ 20___. Contractor ID# _____

Signed _____

Chapter 27 Review
Insulating Structures

Name _____ Date _____ Class _____

Study Chapter 27 of the text, then answer the following questions.

1. List three reasons to install insulation in a building.

   _____
   _____
   _____

_____ 2. The primary type of heat transfer that occurs when a metal panhandle heats up along with the pan is called _____.
   A. radiation
   B. convection
   C. conduction
   D. None of the above.

_____ 3. The primary type of heat transfer that occurs when an object is warmed by the sun is called _____.
   A. radiation
   B. convection
   C. conduction
   D. None of the above.

_____ 4. The primary type of heat transfer that occurs when heat is transferred by the natural movement of warm air is called _____.
   A. radiation
   B. convection
   C. conduction
   D. None of the above.

_____ 5. A number that represents how well a material resists heat movement is known as the _____.

6. Calculate the overall R-value for this wall. Show your work in a neat, orderly fashion.

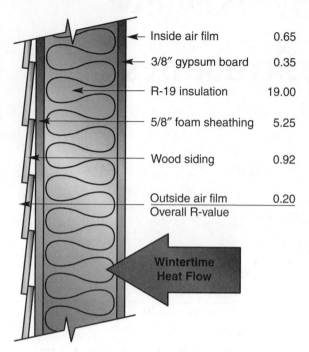

Inside air film	0.65
3/8" gypsum board	0.35
R-19 insulation	19.00
5/8" foam sheathing	5.25
Wood siding	0.92
Outside air film	0.20
Overall R-value	

Wintertime Heat Flow

_____ 7. True or False? *Infiltration* is the flow of air through cracks and gaps in a building that result in unwanted heat loss or heat gain.

_____ 8. Infiltration can be reduced by _____.
   A. weather stripping doors and windows
   B. installing an exhaust fan
   C. installing air conditioning
   D. All of the above.

Match the type of insulation with the correct description

_____ 9. Made from foil that reflects radiant energy.

_____ 10. Clumps of material that are poured or blown into place.

_____ 11. Made of plant material or foam plastic.

_____ 12. Precut to match stud spacing.

A. Rigid
B. Blanket
C. Fiberglass
D. Reflective
E. Loose fill

_____ 13. Sealant is used in or around _____ to reduce infiltration.
   A. doors
   B. window trim
   C. siding joints
   D. All of the above.

Name _____

14. What is a vapor barrier?

_____
_____
_____

_____ 15. True or False? Rigid insulation can be nailed into place.

16. List four safety precautions to follow when working with insulation.

_____
_____
_____
_____

17. Talking and street noise are two examples of _____ sound. Vibrations that move through a structure are known as _____ sound.

_____

18. List three methods of reducing sound transfer.

_____
_____
_____
_____
_____

Name _____ Date _____ Class _____

Activity 27-1
Installing Insulation

In this activity, you will learn to install batt-type blanket insulation. Insulation reduces heating and cooling costs and helps maintain a constant temperature within a building. Recall that insulation is installed after the building exterior is enclosed and rough-in of plumbing, HVAC, electrical power, and communication systems is completed.

Introduction

Insulation installation begins on exterior walls. One installation challenge is fitting insulation around obstacles within the wall. Obstacles can include electrical boxes, wires, and plumbing. Small spaces around window and doorjambs also require special attention. Failure to fill these areas reduces the effectiveness of the insulation. Simply stuffing insulation into these areas also reduces the effectiveness of the installation.

Two solutions to this challenge are to carefully cut and fit small pieces of insulation into the space or to spray foam insulation in those areas. Once problem areas are insulated, the remainder of the stud cavity is filled with insulation. Insulation that has a vapor barrier should be installed with the barrier facing the inside of the building. Barrier edges are secured with staples.

Insulation without a vapor barrier is friction fit. Once installed, a vapor barrier is created by placing a plastic sheet over the entire wall. A plastic sheet vapor barrier can also be attached to the ceiling joists on a top floor for insulation that is blown in after the drywall is installed. This technique provides an excellent vapor barrier, reduces labor cost, and avoids the challenges of doing overhead installations.

Safety Notes

Fibers from fiberglass and rock wool insulation irritate the skin and respiratory tract and can be difficult to remove from eyes. Wear a well-fitted respirator, leather gloves, clothing that covers your body, and safety glasses or goggles. Leather gloves or gloves made from materials the fibers are not likely to penetrate work well.

Tools and Supplies

- Tape measure
- Straightedge
- Utility knife
- 30″ × 30″ plywood cutting surface
- Staple gun
- Blanket insulation

Copyright by The Goodheart-Willcox Co., Inc.

Procedure

1. Determine if it is necessary to fit insulation around any obstacles. If so, cut the insulation and fit it around the obstacles. To reduce thickness, carefully pull away part of the fiber.

2. Measure the width of the stud cavity and the insulation you will use. The insulation should fit snugly in the cavity, without being crushed.

3. Measure the height of the stud cavity and add 1/2".

4. Move the insulation onto the plywood board. Measure the required length. Place the straightedge across the insulation at this point.

5. Compress the insulation with the straightedge. Cut along the straightedge with the utility knife.

6. Repeat Steps 4 and 5 until all exterior wall stud cavities are filled.

7. If there are any narrow stud cavities, cut the insulation to that width. As in Step 4, mark the necessary width with the straightedge. Compress the insulation with the straightedge and cut along the edge with the utility knife. To protect the surface below the cutting surface, continue to move the insulation as needed so that the plywood sheet is always beneath the insulation.

8. Fill areas around window and doorjambs with trimmings.

9. If the insulation does not have a vapor barrier, install one using the method outlined in the introduction.

10. Cover the underside of the ceiling joist with a plastic sheet in preparation for blown insulation.

11. Ask your instructor to check your work.

Chapter 28 Review
Finishing the Building

Name _____ Date _____ Class _____

Study Chapter 28 of the text, then answer the following questions.

_____ 1. Finish work includes installing _____.
  A. sinks and lavatories
  B. drywall
  C. light fixtures
  D. All of the above.

_____ 2. Which of the following statements about plaster is *not* true?
  A. Is a mixture of Portland cement, aggregate, and water.
  B. Can be applied over woven wire mesh, expanded metal, or wood lath.
  C. Covers interior wall and ceiling frames.
  D. Is applied in three layers.

_____ 3. True or False? Drywall is expensive and difficult to install.

4. What is a furring strip?

_____

_____

_____

5. Place the numbers 1–4 in the blanks preceding the steps for finishing drywall joints to indicate the sequence in which they are normally completed.

_____ Rough spots are sanded lightly and a third coat of joint compound is applied.

_____ Joint compound is applied over the joints, followed by a strip of special paper tape that reinforces the joint.

_____ After the compound dries, it is sanded smooth.

_____ After the first layer dries, a second layer of joint compound is added, smoothed, and left to dry.

_____ 6. Temporary walls used to divide open spaces are called _____ walls.

7. List five types of finished flooring.
_____
_____
_____
_____
_____

8. List three materials commonly used as underlayment.
_____
_____
_____

9. Place the numbers 1–6 in the blanks preceding the steps of cabinet installation to indicate the sequence in which they are typically completed.

_____ Check the walls for straightness and plumb. Trim the cabinet or shim low spots so the cabinets will align properly.

_____ Install the base cabinets, following the same procedure.

_____ Install countertops.

_____ Locate and mark wall studs.

_____ Put top cabinets in place and secure them to the studs with screws.

_____ Determine the location of each cabinet unit on the walls.

_____ 10. Doors, stairways, and handrails are installed by _____.

_____ 11. True or False? Water is an ingredient in oil-based paint.

_____ 12. Paint gets its color from _____.
    A. binder
    B. additives
    C. pigment
    D. All of the above.

_____ 13. The substance(s) in paint and varnish that produce(s) the film is known as _____.
    A. binder
    B. additives
    C. pigment
    D. All of the above.

Name _____

_____ 14. Which ingredient in paint and varnish evaporates after the coating is applied?
   A. binder
   B. solvent
   C. pigment
   D. All of the above.

15. Name two material used to cover walls.

_____

_____

16. Channels designed to collect water draining from a roof are called _____. Vertical tubes or pipes that direct water away from the building foundation are called _____.

_____

_____

_____ 17. Finishing the installation of the plumbing system includes _____.
   A. installing fixtures
   B. connecting appliances to supply lines and drains
   C. testing piping systems for leaks
   D. All of the above.

Name _____ Date _____ Class _____

Activity 28-1
Installing Drywall

In this activity, you will learn to install drywall, also called gypsum board. It is often used for ceilings and walls. Ceilings are covered first so the top edge of the panels fastened to the wall can help support the edge of the ceiling along the wall.

When it is necessary to cut openings in drywall, a jab saw is used. The point of the saw is pushed through the panel to begin sawing. Review the following suggestions regarding safe work practices before beginning to install drywall.

Safety Notes

Drywall is heavy and requires two people for installation. Overhead installations are especially challenging. The module's ceiling requires a panel that is approximately 4′ × 4′. Carefully review the procedure and devise a plan before starting the installation. Have all tools and fasteners within reach during installation.

Tools and Supplies

- Tape measure
- Chalk line
- Drywall T-square
- Utility knife
- Jab saw
- Drywall screw gun
- Extension cord
- Nail apron for drywall screws
- Drywall lifter or 12″ flat pry bar
- Seventy five 1 5/8″ drywall screws

Procedure

1. Measure the width and length of the ceiling. Transfer these dimensions to a drywall panel and use a chalk line to layout the lines.

2. Position the straightedge along a chalk line. Using the utility knife and using the straightedge as a guide, cut through the paper covering the surface of the board and into the core approximately 1/8″. Stand the panel on edge. Place your knee against the panel, directly behind the cut made on the opposite surface. Grab one end of the panel and pull to snap the panel along the cut line. Once the panel is snapped, use the utility knife to cut through the paper backing to completely separate the two pieces.

Copyright by The Goodheart-Willcox Co., Inc.

3. Once the drywall panel is in place, you will not be able to locate the ceiling joists. Lightly mark lines on the surface of the drywall to indicate the center of each ceiling joist. Position the panel so you know which edge will be against which wall. Measure the distance from one wall to the center of each end of the first ceiling joist parallel to the wall you are measuring from. Transfer these dimensions to the drywall. Draw a light pencil line connecting the two points. Repeat this process to locate the center of the other ceiling joists.

4. Beginning approximately 2" from the end of each ceiling joist centerline, lightly draw an Xs at 8" intervals along each of the centerlines. These mark the location for each drywall screw.

5. Move a stool or platform in place that is the correct height. The person driving the screws should put on the nail apron with drywall screws in the pocket. Plug in the drywall screw gun and extension cord, then test the gun. Position the screw gun so you can reach it from the stool or ask another class member to hand it to you when needed.

6. Lift the panel, face-side down, and position it against the ceiling joists. The person holding should use both hands so the panel does not shift. The person driving the first screws should use their head to hold their portion of the panel firmly against the ceiling joists.

7. Place a drywall screw on the drywall screw gun and drive it in at an X. Drive three or four more screws before you move your head. With your section secured, drive several screws near your classmate who is holding the panel. This should be enough screws to hold the panel in place. Have your classmate drive the remaining screws.

8. Drywall is run horizontally on walls, so the long edges of a drywall panel are tapered to provide space for taping the joints. Two tapered edges should butt together at each horizontal wall joint. Measure and cut a panel to fit on one wall against the ceiling.

9. Measure and transfer dimensions for openings for windows or electrical boxes. Double check the dimensions before making any cuts. Use the jab saw to cut out the openings.

10. Lightly mark the centerline of each stud on the ceiling. Since you will be able to see each stud below the panel, these marks should be sufficient to position the screws for the top wall panel. Similar marks on the floor will help locate the studs behind the lower panel. Make sure you have the tools and supplies you need before lifting the panel.

11. With the help of a classmate, lift the panel into place. Drive one screw through the top of the drywall and into each stud. Do the same at the bottom of the panel. This will hold the panel in place, allowing your classmate to drive the remaining screws.

12. Repeat Steps 8 through 11 to install the remaining drywall. Cut the bottom panels 1/4" shorter than the opening is tall. In this way, the panels can be positioned and lifted tight against the lower edge of the top panel. Use a drywall lifter or a 12" flat pry bar to raise and hold the panel in place until screws can be driven into place.

13. Store usable pieces of drywall as directed by your instructor. Return tools and supplies to storage and clean the work area.

14. Ask your instructor to check your work.

Name _____ Date _____ Class _____

Activity 28-2
Painting

Painting improves the appearance of many surfaces and protects the base materials. Surface preparation, priming, and topcoat application are the major steps in the painting process.

This is a two-part activity: Part 1 focuses on exterior painting, Part 2 focuses on interior painting. Review the following safety notes before you begin this activity.

Safety Notes

You will be using water-based paints in this activity. Even though they do not contain the amount of solvents found in oil-based paints, ventilation is still important. Even small amounts of solvent can irritate some individuals.

Wear disposable plastic gloves to keep paint off your hands and make cleaning your hands less time-consuming. Wear a jacket or old shirt over your clothing to protect it from paint drips and splatters.

You will need to climb a ladder to reach trim on the exterior of the module. Ensure that the legs of the stepladder are firmly placed on solid ground. Do not place anything loose under the legs. Get assistance from your instructor if you have trouble positioning the ladder so it is stable.

Tools and Supplies

- 6′ stepladder
- Nail set
- Hammer
- Caulking gun and caulk
- 1″ putty knife
- Sanding block
- Duster
- 2″ paintbrush
- Paint roller with cover
- Extension handle for paint roller
- Paint roller tray
- Drop cloth
- Exterior filler
- Spackle
- Abrasive paper
- Sealer for knots

Copyright by The Goodheart-Willcox Co., Inc.

- Exterior house paint
- Interior wall paint
- Paint paddle
- Rag

Procedure

Part 1—Exterior

_____ 1. Examine the surfaces to be painted for finish nails that have not been properly set. Drive these nails 1/8" below the surface using the nail set and hammer.

_____ 2. Spread a drop cloth or other protective cover over the lab floor. Set up the stepladder and ensure it is stable.

_____ 3. Look for surface defects that need to be filled. Use the putty knife to place exterior filler in the defects and in any holes above set nails. Clean the putty knife when you are finished.

_____ 4. Check the joints between the siding and trim. If they have not been caulked, apply a small bead of caulk along the joint. Then press it into the joint using your index finger.

_____ 5. Allow the filler and the caulk to cure several hours before proceeding. Return tools and supplies to storage and clean the work area.

_____ 6. Sand the filler flush with the surrounding surface. Do not sand the caulk. Use the duster to remove dust from all of the surfaces to be painted. Dust the surfaces again with a rag.

◯ 7. Ask your instructor to check your work and initial the circle.

_____ 8. If the surface to be painted includes knots, apply a coat of sealer before applying primer. Sealer reduces the chance that sap in the knots will bleed through the paint.

_____ 9. Use a rag or a small foam brush to apply the sealer. Clean the brush or rag. Allow the sealer to dry for the time noted on the container directions.

_____ 10. Open the exterior primer. Stir it carefully and thoroughly. Ensure that it is completely mixed. If directed to do so, transfer a small amount of primer to a separate container for your use and pass the larger container on to other students.

_____ 11. Begin applying primer at the highest part of the wall and work downward. Dip the brush into the primer and tap the bristle on the inside of the container to prevent dripping. Begin each brush stroke on an uncoated part of the surface and brush toward the part that has been coated. When painting horizontal siding, paint the bottom edge before painting the surface of each strip of siding. Clean the brush and put away the primer when you are finished.

_____ 12. Once the primer is dry, repeat Steps 10 and 11, applying a coat of exterior paint. If time permits, your instructor may want you to apply a second coat.

◯ 13. Ask your instructor to check your work and initial the circle to the left.

Name _____

Part 2—Interior

_____ 14. Examine the surfaces to be painted for finish nails that have not been properly set. Drive these nails 1/8″ below the surface using the nail set and hammer.

_____ 15. Spread a drop cloth or other protective cover over the floor.

_____ 16. Look for surface defects that need to be filled. Use the putty knife to place spackle in the defects and in any holes above set nails. Clean the putty knife when you are finished.

_____ 17. Check the joints at the corners of window trim and where baseboards meet at the inside corner of the module. If necessary, apply caulk to fill these joints.

_____ 18. Let the spackle cure and then sand it flush with the surrounding surface. The caulk should be cured by the time the upper part of the walls are painted and can be painted at that time.

( ) 19. Ask your instructor to check your work and initial the circle.

_____ 20. Open the interior primer. Stir it carefully and thoroughly. Ensure that it is completely mixed. If directed to do so, transfer a small amount of paint to a separate container for your use and pass the larger container on to other students.

_____ 21. Begin painting the ceiling. Use a brush to apply paint at the corners where the ceiling meets the walls. Use the roller to coat the remainder of the ceiling. Attach an extension pole to the roller frame to eliminate the need for a ladder or stool.

_____ 22. Paint window trim and baseboard. Clean the brush and roller and put away the tools and supplies. Ask your instructor to check your work before you continue.

_____ 23. Once the primer is dry, repeat Steps 21 and 22, applying a coat of interior paint. If time permits, your instructor may want you to apply a second coat.

( ) 24. Ask your instructor to check your work and initial the circle.

Copyright by The Goodheart-Willcox Co., Inc.

Chapter 29 Review
Landscaping

Name _____ Date _____ Class _____

Study Chapter 29 of the text, then answer the following questions.

_____ 1. The process of _____ an outdoor area is known as landscaping.
   A. designing
   B. modifying
   C. maintaining
   D. All of the above.

Identify each of the following elements as a type of softscape (S) or hardscape (H).

_____ 2. Concrete

_____ 3. Bricks

_____ 4. Ground cover

_____ 5. Flowers

_____ 6. Mulch

_____ 7. Gravel

_____ 8. Sod

_____ 9. Asphalt

_____ 10. Deciduous trees

_____ 11. Pressure-treated wood

_____ 12. Coniferous trees

_____ 13. Stone

14. Place the numbers 1–5 in the blanks preceding the steps in the landscaping process to indicate the sequence in which they are typically completed.

    _____ Erecting features such as walkways, retaining walls, and driveways.

    _____ Maintaining the site.

    _____ Planting trees, shrubs, flowers, and other plants.

    _____ Cultivating the soil.

    _____ Contouring the site.

15. The _____ shows the type and placement of all plants, hardscape features, and the finished contour of the land.

    _____

    _____ 16. True or False? Topsoil contains very little organic matter.

    _____ 17. Items added to an outdoor site that are not part of the main structure are called _____.

18. Define *cultivation*.

    _____

    _____

19. Planting is done in four steps. List these steps in the order they are typically completed.

    _____

    _____

    _____

    _____

    _____

    _____

    _____ 20. Mulch is a protective layer of material used to _____.

    A. retain moisture

    B. decrease erosion

    C. suppress weed growth

    D. All of the above.

21. List two methods for planting a lawn.

    _____

    _____

22. State two reasons why landscape maintenance is important.

    _____

    _____

    _____

Name _____ Date _____ Class _____

Activity 29-1
Landscaping—Planting

One of the most labor-intensive tasks in landscaping is transplanting large plants, such as trees. These plants have root balls contained in either pots or burlap fabric. These large plants can be difficult to move and require large holes for planting. Landscapers follow the landscape plan to determine plant location.

In this activity, you will plant either a tree or shrub. Your instructor will show you where you are to plant it.

Tools and Supplies

- Rule
- Spade
- Rake
- Wire cutters
- Pliers
- Hammer
- Two-gallon bucket
- One tree or shrub
- Two gallons of peat moss
- Two quarts of general purpose fertilizer
- Two 2″ × 2″ × 6′ stakes
- Two, one-foot lengths of garden hose
- Six feet of wire
- Five gallons of water
- Five gallons of mulch

Procedure

_____ 1. Dig a hole that is twice the diameter of and equal in depth to the root ball on the plant.

_____ 2. Add peat moss and fertilizer to the soil that was removed. The amount of peat and fertilizer needed varies based on the type of soil type in the planting area. A basic formula for amended soil, however, is one shovel of peat moss and one cup of general purpose fertilizer for every four shovels of soil.

Copyright by The Goodheart-Willcox Co., Inc.

_____ 3. Place the plant in the center of the hole. Add amended soil to the hole and around the plant to make it stable.

_____ 4. Look at the plant. Is it vertical? Is the top of the root ball level with or slightly higher than the undisturbed ground around it? Adjust the plant as needed. When you are satisfied with the position of the plant, compress the soil that was added in Step 3.

_____ 5. Add water around the root ball. Add more soil to fill the remaining space around the root ball. Compress the soil again and add more water. Once again, check the position of the plant. Adjust by adding and compressing additional soil.

_____ 6. Finish filling the hole. Form a rim in the soil around the perimeter of the excavated hole. This will help the water remain in place while it soaks into the soil.

_____ 7. If necessary, stake the tree as shown here.

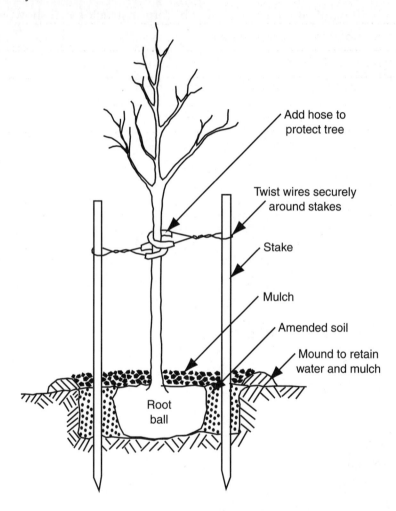

Cover the wire that touches the plant with hose to prevent damage to the plant.

_____ 8. Place at least 3" of mulch around the base of the plant. This will help the soil hold moisture and reduce the growth of weeds.

○ 9. Have your instructor check your work and initial the circle at the left.

_____ 10. Rake the area. Remove any debris from the site. Return tools and supplies to storage.

Name _____ Date _____ Class _____

Activity 29-2
Sustainable Landscaping

Sustainable landscaping uses materials that minimize the need for irrigation, grow well in the local environment, are disease resistant, and require minimal use of fertilizer and other chemicals. This assignment focuses on landscaping materials that meet these criteria for your area.

With your instructor's help, select one of the following topics about sustainable materials. Prepare a report, presentation, or display that describes the landscape materials you researched and outline those that are most appropriate for residential use in your area.

A reliable source of information about landscaping is the extension service of your state university's college of agriculture. Search the Internet for *extension university landscaping* or *extension university horticulture*.

Possible topics:

- Mulch, decorative rock, and other nonliving materials.
- Grass.
- Ground cover.
- Deciduous shrubs or trees.
- Coniferous shrubs or trees.
- Other topic as approved by your instructor.

Your instructor will provide specifics regarding your report such as length, presentation time, and display size.

Chapter 30 Review
Final Inspection, Contract Closing, and Project Transfer

Name _____ **Date** _____ **Class** _____

Study Chapter 30 of the text, then answer the following questions.

1. List six items a project contract addresses.

   _____
   _____
   _____
   _____
   _____
   _____
   _____
   _____

_____ 2. True or False? The only time inspections are made is at the completion of a project.

_____ 3. Qualified inspectors are _____ in their fields.

4. Briefly describe the three main inspection categories that are reviewed when inspections are made.

   _____
   _____
   _____
   _____
   _____
   _____

5. Name the five steps in a final inspection in the order they are typically completed.

   _____
   _____
   _____
   _____
   _____

6. A(n) _____ identifies problems that need to be fixed before the owner will make final payment on the contract.

_____ 7. True or False? Inspectors who complete the final inspection sign the certificate of completion.

_____ 8. Documents that confirm the items on the punch list have been corrected are called _____.
   A. certificates of substantial completion
   B. certificates of completion
   C. approval forms
   D. payment bonds

_____ 9. During the contract closing, the contractor provides signed _____ to the owner.
   A. approvals
   B. releases
   C. warranties and manuals
   D. All of the above.

_____ 10. True or False? A lien is a written record of the transfer of ownership from one person to another.

_____ 11. The person who brings legal action is called the _____.

12. Define *defendant*.

_____
_____
_____

_____ 13. A (n) _____ is a document guaranteeing that there are no defects to a product.
   A. warranty
   B. guarantee
   C. deed
   D. title

_____ 14. True or False? Foreclosure is a legal process that transfers ownership of the property to the contractor if the client fails to pay.

15. When is a contract considered completed?

_____
_____
_____
_____

Name _____ Date _____ Class _____

Activity 30-1
Transferring the Project

Transferring ownership of a construction project from contractor to owner involves several steps. When a contractor nears completion of the project, the owner is asked to perform an inspection and prepare a punch list identifying things that need to be corrected before the transfer takes place. Once the contractors correct the problems, the transfer process proceeds.

In this activity, your instructor will act as the owner of the structure. All forms are provided at the end of this activity.

1. Ask your instructor to conduct an inspection and then complete the punch list based on that inspection.

2. If time permits, have the contractors make the needed repairs and the instructor complete a final inspection. If there is no time for repairs, act as if the repairs have been made and move on to Step 3.

3. Once the owner is satisfied that the structure is acceptable, ask the contractors to sign the Contractor Warranty Form. In a real project transfer, this is also the time that the contractors would provide copies of warranties and manuals to the owner.

4. The contractors now sign a Release Form that states the owner will not be liable for claims made by any suppliers or workers involved with the project. This assures the owner that all contractor expenses have been paid in full by the contractors.

5. It is now time to close the contract. The owner (your instructor) will make a final payment (award grades) and assume full responsibility for the module.

Name _____

Punch List
Exterior
Roof including overhang
Front wall
Side wall
Interior
Ceiling
Walls
Floor
Plumbing system
Electrical system
HVAC system
Communication system

Copyright by The Goodheart-Willcox Co., Inc.

Name _____

Contractor's Warranty Form

We the undersigned, contractors for Module _____, do hereby guarantee the quality of work and materials used in the construction of this module for a period of one year, beginning _____.
Contractor signatures:

_____
_____
_____
_____
_____
_____
_____
_____
_____
_____
_____
_____
_____
_____
_____
_____
_____
_____
_____
_____
_____
_____
_____

Name _____

Contractor's Release Form

We the undersigned, contractors for Module _____, do hereby release all claims and liens against the module constructed for this course during _____ school year.

Contractor signatures:

_____
_____
_____
_____
_____
_____
_____
_____
_____
_____
_____
_____
_____
_____
_____
_____
_____
_____
_____
_____
_____
_____
_____
_____

Chapter 31 Review
Project Operation, Maintenance, and Repair

Name _____ Date _____ Class _____

Study Chapter 31 of the text, then answer the following questions.

Match the description of the task with the correct term.

_____ 1. Done as part of normal use.

_____ 2. Done periodically to retain the original functionality of a device.

_____ 3. Done as needed on malfunctioning devices to return them to functionality.

A. Maintenance
B. Rehabilitation
C. Operation
D. Repair

Identify each of the following tasks as an operation (O), maintenance (M), or repair (R).

_____ 4. Testing smoke alarms.

_____ 5. Replacing the cord on a desk lamp.

_____ 6. Removing snow from the driveway.

_____ 7. Setting the timer for an irrigation system.

_____ 8. Replacing a defective component in the security system.

_____ 9. Disassembling, cleaning, lubricating, and reassembling a fan.

_____ 10. Cleaning restrooms.

_____ 11. Repainting a foyer.

12. Name two types of property protection systems.
    _____
    _____

_____ 13. Project _____ involves daily activities that preserve the optimal appearance and functionality of a property.

14. List five examples of housekeeping tasks.
    _____
    _____
    _____
    _____
    _____

15. Distinguish between periodic maintenance and maintenance-as-needed.
    _____
    _____
    _____
    _____

_____ 16. True or False? The decision to rebuild or replace a malfunctioning component is made using a problem-solving approach.

_____ 17. Returning something to its original state is called _____.
    A. replacing
    B. regeneration
    C. rebuilding
    D. reassembly

Name _____ Date _____ Class _____

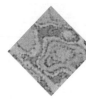

Activity 31-1
Project Operation, Maintenance, and Repair (OMR) Calendar

The purpose of this activity is to develop a monthly list of tasks that need to be done at your home to keep it functioning properly. List all tasks, regardless of who does the actual work. The following examples will help you get started.

Operational Tasks

- Check for water leaks at toilets, faucets, and showerheads.
- Switch HVAC control from heating and cooling.
- Activate irrigation system.
- Remove garbage.
- Check landscaping for damage and disease.

Maintenance Tasks

- Interior cleaning (may divide into areas to allow for varying frequency).
- Trim trees and shrubs.
- Change filters in HVAC system.
- Lubricate hinges and locks.
- Fertilize plants.

Repair Tasks

- Address problems found during inspections.
- Touch-up paint.
- Replace lightbulbs.
- Replace defective HVAC system parts.

Use the table found at the end of this activity to list the tasks. If a task must be done more than monthly, indicate the number of times as a part of the description.

Copyright by The Goodheart-Willcox Co., Inc.

OMR Calendar

Month	Tasks		
	Operation	Maintenance	Repair
January			
February			
March			
April			
May			
June			
July			
August			
September			
October			
November			
December			

Name _____ Date _____ Class _____

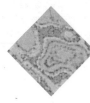

Activity 31-2
Drywall Repair

Drywall repair is a common home repair project. Damaged areas can be small or large dents, cracks, or holes. Small areas of damage are repaired using spackle and a putty knife. Large areas of damage require more complex repairs involving joint compound, joint tape, and multiple episodes of sanding.

In this activity, you will repair a hole in the drywall. This requires fitting a plug into the hole. Blocking installed behind the drywall is used to support the plug, shown here.

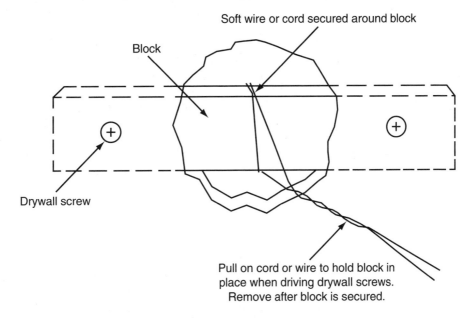

Your instructor will identify the area you will repair.

Tools and Supplies

- Drywall screwdriver
- Taping knife
- Sanding block
- Plywood block, at least 1/2" thick
- Cord or wire, 18"
- Construction adhesive
- Three to six 1 5/8" drywall screws
- One piece of drywall, large enough to make plug
- Joint compound
- Joint tape
- 100 grit abrasive paper

Copyright by The Goodheart-Willcox Co., Inc.

Procedure

1. Examine the plywood block. It should be long enough to extend approximately 2″ beyond either side of the hole and narrow enough to be inserted through the hole.

2. Tie a piece of cord or soft wire around the center of the block. You will use this to pull the block against the back of the drywall while screws are driven.

3. Place a small amount of construction adhesive on the areas of the block that will make contact with the back of the drywall panel. Immediately insert the block through the hole and use the cord or wire to pull it firmly against the back of the drywall. Then drive at least two drywall screws through the drywall into the plywood.

4. Remove the cord or wire and allow the adhesive to cure.

5. Cut a plug of drywall to fit the opening. Secure it to the block with at least one drywall screw.

6. Sand the surface to remove bumps around the plug and screws. Then apply the first coat of joint compound and joint tape. Allow to cure.

7. Sand the area to remove any bumps or high spots. Apply the second coat of joint compound. Feather this coat away from the repair, spreading the compound until there is nothing left to spread. It should extend several inches beyond the repair. Allow to cure.

8. Sand the area again. Apply a third coat of joint compound, feathering the edges as you did in Step 7. Allow to cure.

9. Sand the area one more time. Examine the finished surface for defects. Fix any defects that remain by applying compound and sanding. Once you are satisfied with the repair work, the surface is ready for paint.

Chapter 32 Review
Remodeling Buildings

Name _____ Date _____ Class _____

Study Chapter 32 of the text, then answer the following questions.

1. Limited cosmetic change to an existing structure is called _____. Altering the structure of a building is called _____.

   _____

   _____

Classify each of the following tasks as either remodeling (RM) or redecorating (RD).

_____ 2. Adding a room to a home.

_____ 3. Repainting a living room.

_____ 4. Adding a deck to the rear of a home.

_____ 5. Replacing carpet.

_____ 6. Wallpapering a living room.

_____ 7. Making a kitchen handicapped accessible.

_____ 8. Converting an unfinished basement into a family room and storage room.

_____ 9. True or False? The remodeling process is essentially the same as the process for planning and constructing a new building.

10. The four basic tasks that need to be done when preparing a remodeling site are:

    _____

    _____

    _____

    _____

_____ 11. Project alteration includes _____ components.
   A. adding
   B. removing
   C. relocating
   D. All of the above.

12. Why it is important during demolition work to wear eye protection and clothes that cover the skin?
    _____
    _____
    _____
    _____

_____ 13. True or False? Electricity and water should be left on when walls are removed.

_____ 14. True or False? Eye protection, dust mask or respirator, gloves, and safety shoes should be worn when removing drywall.

15. List three items commonly removed during a kitchen remodel that *can* be reused or recycled.
    _____
    _____
    _____

16. List three items commonly removed during a kitchen remodel that *cannot* be reused or recycled.
    _____
    _____
    _____

17. Name five things that are likely to be replaced when a kitchen is remodeled.
    _____
    _____
    _____
    _____
    _____

_____ 18. The objective of _____ is to return an older building to its original condition.

Name _____ Date _____ Class _____

Activity 32-1
Room Remodeling Project

In this activity, you will plan a remodeling project. The activity is divided into four parts. Your instructor will review your progress and provide feedback at appropriate times. Read the entire assignment before you begin working on Part 1.

Part 1 _____

Date due _____

The purpose of Part 1 is to identify a room you think needs to be remodeled. What needs will be met as a result of the remodeling? You will prepare a scale drawing of the room that identifies all items that require floor space.

1. Identify the room you would like to remodel. Consider a room that is currently not useful, such as a kitchen that is not accessible for its physically disabled owner. The room does not need to be in your home, but it must be a real room. You must have access to the room and be able to talk to the residents regarding their needs.

2. Answer the following questions on a separate piece of paper. Then prepare a scale drawing of the room using the graph paper labeled "Floor Plan A" at the end of this activity. Use the largest scale that will fit on the paper. Show all windows, doors, and accessible closets. Include overall room dimensions and the dimensions of any offsets in the walls. Label what is located on the opposite side of each wall.

    A. What type of room will this be once it is remodeled (bedroom, office, bath)?

    B. Describe the needs that will be served by the remodeled room.

    C. Identify furniture, cabinets, appliances, and other items that will require floor space in the remodeled room.

Part 2 _____

Date due _____

In this part of the activity, you will consider alternative locations for the furniture, cabinets, appliances, and other items that need to be accommodated in the remodeled room.

1. Review the list you created in Part 1 of needs and items to be included in the remodeled room. Then answer the following questions on a separate piece of paper.

    A. Does anything need to be added? If so, what?

    B. Can anything be eliminated? If so, what?

    C. Can any items be combined to reduce floor space requirements? If so, what?

Copyright by The Goodheart-Willcox Co., Inc.

D. Can any of these items be placed in a closet? If so, what?

E. What are the dimensions of the floor space required by each essential item?

2. Draw and label templates of each item that requires floor space. Use the graph paper labeled "Templates" found at the end of this activity. Draw them to the same scale as the floor plan prepared for Part 1, so the space they occupy is an accurate representation of each item.

3. Cut out the templates. Experiment with various layouts using the floor plan you created in Part 1. Answer the following questions on a separate piece of paper.

   A. Did you find a layout that works well within the existing room? If yes, draw this layout on the floor plan you drew. Label each item.

   B. If you did not find a workable floor plan, can any items be omitted? If yes, what? Is it now possible to develop a workable floor plan? If yes, draw this layout on the floor plan developed in Part 1.

   C. Is additional space still needed? Can an interior wall be moved to provide the needed space, without seriously damaging the room from which the space is taken? If yes, which wall? Why does moving this wall work? Draw a revised floor plan showing the location of the newly relocated wall on the graph paper labeled "Floor Plan B" at the end of this activity. Draw and label the location of each of the items.

   D. If moving an interior wall is not workable, can an exterior wall be moved to gain the needed space? If yes, revise the floor plan you created in Part 1 on the graph paper labeled "Floor Plan B." Showing the location of the relocated wall, draw and label the location of each item.

   E. If you cannot create a functional plan to accommodate the needed items, carefully reconsider the items to be included in the room. Can anything be eliminated? If yes, what? How much floor space will be saved?

   F. Can anything be positioned above something else? If yes, what and how much floor space will be saved?

   G. Can anything be positioned under something else? If yes, what and how much floor space will be saved?

   H. Revise the templates to account for the changes made in E, F, and G. If everything now fits within the available space, draw and label the location of each of the templates on the floor plan you created in Part 1.

   I. If all the items still cannot be arranged to fit within the available space, begin eliminating items based on their perceived necessity, until a reasonable layout is achieved. Draw and label each of the templates on the floor plan prepared in Part 1.

4. Submit your answers and your final floor plan to your instructor.

Name _____

Part 3 _____

Date due _____

In Part 3, prepare an initial list of the tasks to complete during the remodeling project. The following questions will guide you in preparing the list. Prepare your list on separate paper and submit on the date due.

1. Will excavation be necessary for the new footing and foundation wall?
2. Will the floor frame need to be extended?
3. Will new walls need to be constructed?
4. Will additional roof framing be required?
5. Will protection of adjacent areas be necessary during demolition or construction?
6. What will need to be removed?
7. Will plumbing rough-in require modification?
8. Will changes in the electrical or communication wiring be necessary?
9. Will changes in the HVAC system be necessary?
10. Will windows or doors need to be installed?
11. Will wall and ceiling materials be new?
12. Will new flooring be installed?
13. Will new cabinets be installed?
14. Will trim be required?
15. Will plumbing fixtures be installed?
16. How will the wall be finished?
17. Will light fixtures and communication devices be installed?

Name _____

Drawn by:	Date:	Drawing:	Project:
		Floor Plan A	

Name _____

Drawn by:	Date:	Drawing:	Project:
		Floor Plan B	

Name _____

Drawn by:	Date:	Drawing:	Project:
		Templates	

Chapter 33 Review
Dam Construction

Name _____ Date _____ Class _____

Study Chapter 33 of the text, then answer the following questions.

_____ 1. A barrier that contains a body of water is called a(n) _____.

_____ 2. A dam project that serves more than one purpose is called a(n) _____ dam project.

_____ 3. True or False? A reservoir that provides a city's water must hold enough water to meet the needs of the city during periods of drought.

_____ 4. Devices used to regulate flow of water downstream from a dam and control the water level in a reservoir are known as _____.
   A. public works
   B. abutments
   C. outlet works
   D. water works

5. Identify three reasons to maintain a steady flow of water downstream from a major dam.
   _____
   _____
   _____

6. How does a dam prevent soil erosion?
   _____
   _____

_____ 7. True or False? Water pressure increases based on the depth of the water in the reservoir.

_____ 8. An electrical generating plant that is powered by _____ is called a hydroelectric plant.

9. Identify four factors that are considered when evaluating the environmental impact of major dam construction.

_____
_____
_____
_____

10. Name three individuals or groups who initiate dam construction projects.

_____
_____
_____

_____ 11. Civil engineers design dams with assistance from _____.
   A. mechanical engineers
   B. electrical engineers
   C. agronomists
   D. All of the above.

12. What is a joint venture?

_____
_____
_____

_____ 13. True or False? The first step in planning a large dam and reservoir project is to obtain financing.

14. Name three factors that may impact the cost estimate for a dam.

_____
_____
_____

_____ 15. True or False? Environmental impact statements are required by federal law for all dam construction projects.

Name _____

16. Identify the components of the dam shown here.

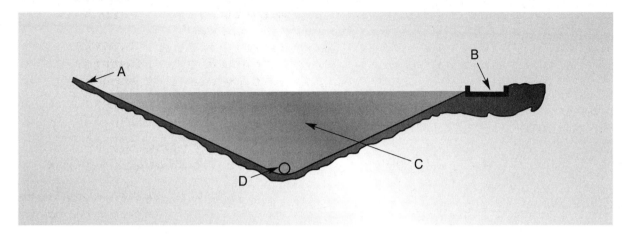

A. _____

B. _____

C. _____

D. _____

_____ 17. Which of the following statements regarding land acquisition is *not* true?

   A. Land for the reservoir and for road relocation is purchased as needed.
   B. Land for the dam and access roads is purchased first.
   C. Land acquisition for a large dam can take several years.
   D. The power of eminent domain cannot be used to acquire the land of unwilling sellers.

18. Define *abutments*.

_____

_____

_____

_____ 19. True or False? Cofferdams are constructed in the area where the outlet works are to be constructed to prevent flooding of the work site.

20. What task is commonly the last done on dam construction?

_____

Name _____ Date _____ Class _____

Activity 33-1
Commercial, Industrial, and Engineered Construction

In this activity, you will prepare a report about a noteworthy major construction topic in commercial, industrial, or engineered construction. Your instructor will indicate the type of report to be prepared and provide guidelines for content and length. Read through the suggested topics for ideas. Look for information on the Internet regarding the topics that interest you.

Bridges

- San Francisco Bay Bridge
- Oakland Bay Bridge
- Tacoma Narrows Bridge
- Sunshine Skyway Bridge
- Akashi Kaikyo Bridge

Dams

- Hoover Dam
- Grand Coulee Dam
- Chief Joseph Dam
- Bath County Pumped Storage Station
- Nurek Dam

Pipelines

- Keystone pipeline
- Gas pipeline system
- Petroleum pipeline system
- Water pipeline system

Roads and Tunnels

- Alaska Highway
- Interstate Highway System
- Eisenhower Tunnel
- New York Third Water Tunnel
- Central Artery Tunnel Project (Big Dig)

Copyright by The Goodheart-Willcox Co., Inc.

Skyscrapers _____

- Empire State Building
- Chrysler Building
- Willis Tower
- Taipei 101
- Otis Elevator history

Sustainable Construction _____

- Earth-sheltered design
- Active solar home design
- Water heating systems
- Green roof systems
- Industrial wind turbines or wind farms

Other Structures _____

- CN Tower
- Seattle Space Needle
- Gateway Arch
- Washington Monument
- NASA Vehicle Assembly Building

1. Select three topics that interest you. You may choose topics not on this list. Briefly describe your three favorite topics on a separate piece of paper and submit your list to your instructor.

2. Your instructor will tell you which topic to do. You may be asked to limit your report or modify the topic slightly to avoid duplications between reports.

   A. What is the topic of your report?
   _____

   B. What limits were suggested?
   _____

   C. What type of report are you to prepare?
   _____

   D. Was a time or length limit suggested? If so, what is it?
   _____

   E. Are you expected to provide a progress report? If so, when?
   _____

   F. When will you present your report?
   _____

Name _____

3. Conduct Internet research on your topic. These suggestions may help you find information:
   - Add *construction* or *design* to the name of the project for more results.
   - Visit related searches suggested by your search engine.
   - Look for references and links in the articles you have found.

4. Create an outline of the information you've gathered and evaluate it.

5. If you need additional information, ask a librarian at your school or at your local public library for help.

6. Refine your outline.

7. Create a first draft of your report. Drafts for visual reports should include a sketch of the display that shows the major elements and how these elements will be linked. Also include a one-page description of each of the major elements of the display. These descriptions may be labeled sketches or outlines.

8. Submit your draft to your instructor for review and approval.

9. Make the revisions suggested by your instructor and create your final report.
   - Review written reports for clarity and for correct spelling, grammar, and punctuation.
   - Time oral reports during practice to ensure they fit into the time limit. Speak clearly and distinctly during practice.
   - Review visual reports for correct spelling, grammar, and punctuation. Check drawings for accuracy. Measure displays to ensure it will fit into the display space.

10. Present your report or display.

Name _____ Date _____ Class _____

Activity 33-2
World's Largest Dams

This assignment focuses on hydroelectric generating facilities. With your instructor's help, select one of the facilities listed below. You may choose another facility with your instructor's approval. Prepare a report, presentation, or display that includes information such as location, construction dates, ownership, builders, material used in construction, electrical generating capacity, volume of water impounded, height and length of dam, provisions for moving water traffic past the dam. Also include information on the benefits (in addition to electrical generation) and problems caused by the dam.

The five largest hydroelectric generating facilities in the world are:

- Three Gorges Dam
- Itaipu Dam
- Guri Dam
- Tucurui Dam
- Grand Coulee Dam

The five largest hydroelectric generating facilities in the United States are:

- Grand Coulee Dam
- Chief Joseph Dam
- Bath County Pumped Storage Station
- Hoover Dam
- The Dalles Dam

An Internet search for any of these facilities will provide many sources of information. If you need additional help, search the Internet for hydroelectricity, world's largest dams, and largest US dams.

Your instructor will provide specifics regarding your report such as length, presentation time, and display size.

Copyright by The Goodheart-Willcox Co., Inc.

Chapter 34 Review
Bridge Construction

Name _____ Date _____ Class _____

Study Chapter 34 of the text, then answer the following questions.

_____ 1. The load that a bridge carries is transferred to the substructure by the _____.

_____ 2. The _____ component of a bridge can include abutments, piers, and footings.

_____ 3. True or False? Piers extend from bedrock to the superstructure.

_____ 4. A bridge that carries traffic when an older bridge is being replaced is called a(n) _____ bridge.
    A. deck
    B. temporary
    C. through
    D. None of the above.

_____ 5. A(n) _____ bridge contains a roadway between trusses.
    A. deck
    B. temporary
    C. through
    D. None of the above.

_____ 6. A bridge constructed with a roadway on top of the main beams or trusses is called a(n) _____ bridge.
    A. deck
    B. temporary
    C. through
    D. None of the above.

Copyright by The Goodheart-Willcox Co., Inc.

7. Name seven types of fixed bridges.

   _____
   _____
   _____
   _____
   _____
   _____
   _____

8. Identify each type of bridge shown in each of the following drawings.

   A. _____

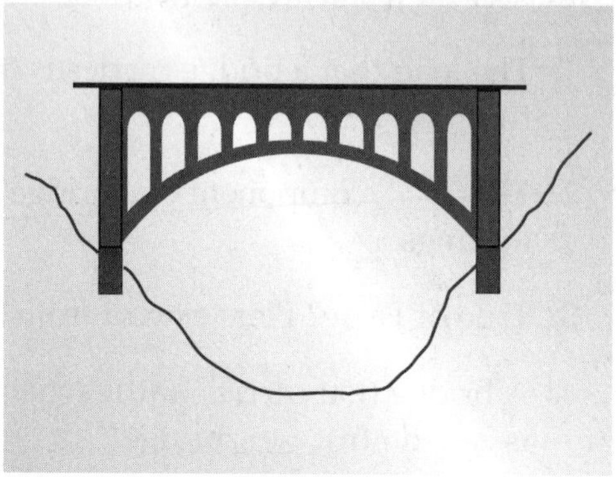

   B. _____

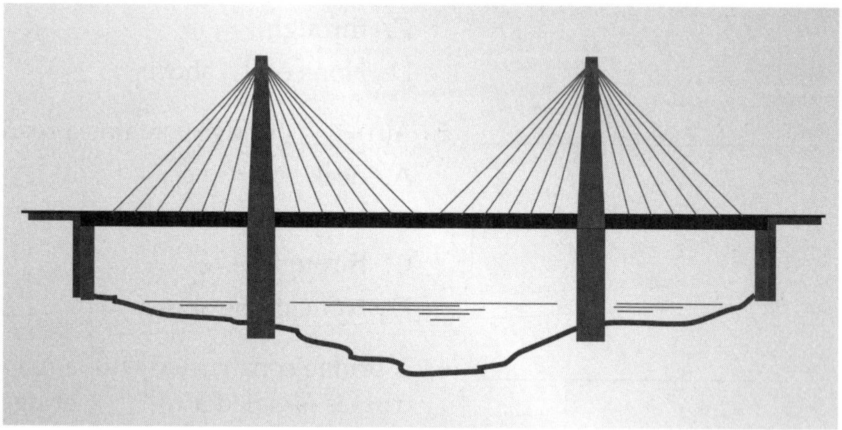

   C. _____

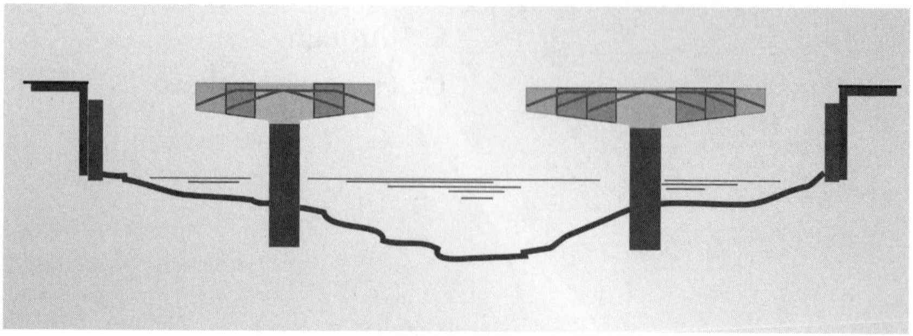

Name _____

D. _____

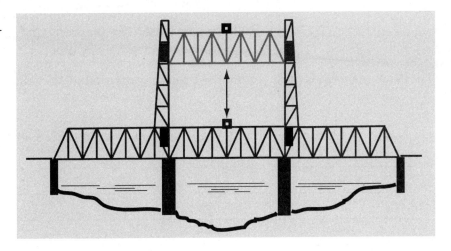

E. _____

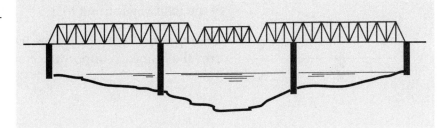

F. _____

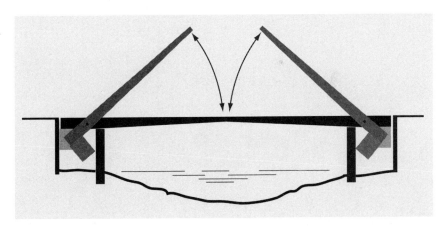

G. _____

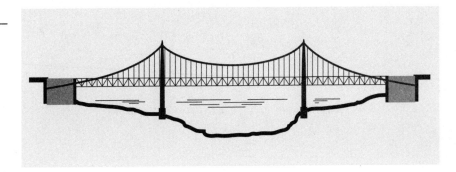

_____ 9. Bridges that rotate on a center pier to allow boat traffic to pass on both sides are called _____ bridges.

_____ 10. True or False? The planning and design phases of bridge building are the same as for any other construction project.

11. What material is most commonly used in construction of bridge substructures?
_____

12. Name four materials commonly used to construct bridge superstructures.
_____
_____
_____
_____

_____ 13. True or False? Bridge construction often begins before construction of the roadway leading to it.

_____ 14. Casting concrete over a bridge framework requires _____ to be built over the frame to support the concrete.

Chapter 35 Review
Road Construction

Name _____ Date _____ Class _____

Study Chapter 35 of the text, then answer the following questions.

_____ 1. True or False? There are approximately four million miles of highways and streets in the United States.

_____ 2. Transportation planners perform all the following tasks except _____.
   A. conducting transportation and travel surveys
   B. supervising highway construction
   C. evaluating the impact of new developments on the existing transportation system
   D. working with various groups involved in the planning of new transportation facilities

Identify each of the following roads as an arterial (A), collector (C), or service (S).

_____ 3. Connect smaller cities and towns to each other.

_____ 4. Generally have two lanes with curbs and space for on-street parking.

_____ 5. High speed limits and access allowed only through ramps at major intersections.

_____ 6. Provides access to adjacent land such as a home, business, or school.

_____ 7. Typically are four or more lanes wide, include traffic signals, and have higher speed limits than service streets.

_____ 8. Medians of concrete or land divide lanes going in opposite directions.

_____ 9. Provide uninterrupted access between large cities.

_____ 10. Have low speed limits and provide access to collectors.

_____ 11. The purpose of a feasibility study for a street or road is to obtain answers to questions such as _____.
   A. What is the projected vehicles volume for the road in the next five or more years?
   B. What is the most practical route for the road?
   C. How long will it take to complete the project?
   D. All of the above.

12. List five basic considerations when planning and designing roads and streets.

_____
_____
_____
_____
_____

_____ 13. Road _____ is expressed as a percentage.

_____ 14. The crown is the _____ point in the center of a road and is designed to move water off the surface and into ditches and drains.

_____ 15. Which of the following statements about pavement is true?
   A. It is the actual surface on which vehicles travel.
   B. Pavement thickness is determined by the contractor on the job.
   C. Pavement is designed to last approximately 2–5 years.
   D. Concrete is the only material used for pavement.

_____ 16. A(n) _____ defines the limits of the construction site once site work begins.

_____ 17. Both public and worker _____ are major considerations of road construction work.

_____ 18. True or False? Road construction is a linear process.

19. Name four machines typically used to move easily excavated materials.

_____
_____
_____
_____

_____ 20. Rippers are prong-like attachments used to break up _____.

Name _____

21. Identify the parts of the roadway shown here.

   A. _____
   B. _____
   C. _____
   D. _____

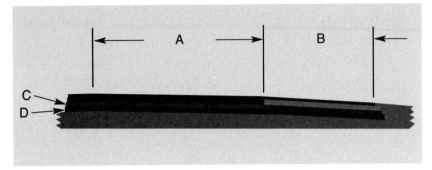

_____ 22. The final task in the road construction process is _____.
   A. erecting signage
   B. placing curbs
   C. landscaping
   D. installing road markings

Chapter 36 Review
Skyscraper Construction

Name _____ Date _____ Class _____

Study Chapter 36 of the text, then answer the following questions.

1. Buildings more than 500 feet tall are considered _____. Buildings taller than 1000 feet are considered _____.

   _____
   _____

   _____ 2. True or False? The development of structural steel and elevators made skyscrapers possible.

3. Briefly state three reasons for building skyscrapers.

   _____
   _____
   _____

4. Place the numbers 1–6 in the blanks preceding the steps in the skyscraper design process to indicate the sequence in which they are typically completed.

   _____ Decision made to proceed or abandon project, based on feasibility study results.

   _____ Design team selected.

   _____ Designers prepare construction documents.

   _____ Design team presents design alternatives.

   _____ Initiator defines the purpose for the skyscraper.

   _____ Design team prepares feasibility study.

5. Who defines the purpose for a building and selects the designers?

   _____

   _____ 6. A(n) _____ is hired to build the skyscraper.

   A. initiator
   B. overseer
   C. superintendent
   D. general contractor

7. Why do skyscraper foundations generally extend to bedrock.

_____
_____
_____

8. What is the purpose of underpinning?

_____
_____
_____

_____ 9. Which of the following is *not* used for demolition work?
   A. Crane fitted with a wrecking ball
   B. Scraper
   C. Cutting torch
   D. Front loader and explosives

_____ 10. A superstructure framework is made up of _____.
   A. beams
   B. girders
   C. columns
   D. All of the above.

Match each piece of equipment with the correct description.

_____ 11. Has a movable horizontal boom that is supported on a fixed tower.

_____ 12. Mounted on tracks that can be moved around a site.

_____ 13. A lifting device that uses a boom and mast guided by a cable or rope.

A. Traveling long-boom crane
B. Tower crane
C. Guy derrick

_____ 14. True or False? Plumbing, HVAC ducts, and elevators can be installed at any time during construction.

Name _____

15. Why must the work be carefully coordinated on a skyscraper construction project?

_____

_____

_____

_____

_____

Chapter 37 Review
Pipeline Construction

Name _____ Date _____ Class _____

Study Chapter 37 of the text, then answer the following questions.

_____ 1. True or False? Gathering systems move natural gas or crude oil from wells to gas processing plants or oil refineries.

2. Define the term *transmission pipeline*.

_____
_____
_____
_____

3. Distribution piping systems deliver _____ and _____ from the pipeline directly to homes, businesses, factories, and other consumers.

_____

_____ 4. Petroleum products are pumped to _____, where they remain until they are moved to their final destination.
   A. tank farms
   B. petroleum towers
   C. underground storage tanks
   D. None of the above.

5. Place the numbers 1–5 in the blanks preceding the following steps in pipeline construction to indicate the sequence in which they are completed.

   _____ Determine route.

   _____ Determine beginning and ending points.

   _____ Obtain easements for buried pipe and purchase land for above ground components.

   _____ Obtain both state and federal permits.

   _____ Determine size of pipe and any branches.

_____ 6. Large valves are located every ____ miles along the main pipeline to control flow.
   A. 2–10
   B. 5–10
   C. 10–20
   D. 5–20

Match the following terms with the correct descriptions.

_____ 7. Used by pipeline companies to monitor and manage the flow of natural gas or petroleum products through the system.

_____ 8. Replacing an area with excavated material.

_____ 9. The process of placing pipes alongside a trench.

_____ 10. Filling a pipeline with pressurized water to test for leaks.

_____ 11. Used to increase pressure in natural gas pipelines to compensate for pressure loss.

_____ 12. System that collects information at metering stations and transmits it to centralized control stations.

_____ 13. Placing a pipe in a trench.

_____ 14. Used to increase the pressure in petroleum and water pipelines to compensate for pressure loss.

A. Directional drilling
B. SCADA system
C. Hydrostatic testing
D. Metering station
E. Depositing
F. Backfilling
G. Stringing
H. Pumping station
I. Compressor station

15. The alternative to the open cut method for installing a pipeline below a street or stream is known as _____.

_____

Chapter 38 Review
Careers in Construction

Name _____ Date _____ Class _____

Study Chapter 38 of the text, then answer the following questions.

1. List six areas of employment directly related to construction.

   _____
   _____
   _____
   _____
   _____
   _____

_____ 2. True or False? Construction education may begin with high school courses in construction, technical drawing, and various trades.

_____ 3. Postsecondary programs in the construction trades are available at _____.
   A. technical schools
   B. community colleges
   C. universities
   D. All of the above.

_____ 4. _____ programs provide on-the-job training with classroom instruction.

Match the type of worker with the correct definition.

_____ 5. A person who oversees the work of journeymen, apprentices, and helpers for a specific trade.

_____ 6. A person with no previous experience as a construction worker.

_____ 7. A person who calculates the cost of completing a specific job.

_____ 8. A person qualified to work with little supervision and to have a helper or apprentice assist.

_____ 9. A person who oversees all of the site work for a particular job.

_____ 10. A helper enrolled in a training program that includes classroom instruction.

A. Helper
B. Apprentice
C. Journeyman
D. Master tradesperson
E. Supervisor
F. Superintendent
G. Estimator

_____ 11. Foundational skills include _____.
   A. basic skills
   B. thinking skills
   C. personal qualities
   D. All of the above.

_____ 12. Which of the following is *not* one of the basic skills for workplace success?
   A. Reading
   B. Writing
   C. Decision making
   D. Listening

13. List the five personal qualities essential for successful employment.

_____
_____
_____
_____
_____

_____ 14. True or False? To be competent means having the knowledge to apply skills and use available technologies productively.

Name _____

15. Write a brief description for each area of competency that workers are expected to possess.

   A. Utilizing resources

   _____
   _____
   _____

   B. Interpersonal skills

   _____
   _____
   _____

   C. Information processing skills

   _____
   _____
   _____

   D. Systems

   _____
   _____
   _____

   E. Technology

   _____
   _____
   _____

16. Place the numbers 1–6 in the blanks preceding the steps in getting a construction job to indicate the sequence in which they are typically completed.

   _____ Begin contacting prospective employers.

   _____ Schedule interview.

   _____ Identify employment opportunities.

   _____ Decide what type of construction work you want to do.

   _____ Accept or reject a job offer.

   _____ Prepare a brief résumé.

17. List five items that are normally reviewed in a job evaluation.

   _____
   _____
   _____
   _____
   _____

_____ 18. A positive progression is also known as a(n) _____.

_____ 19. An employee who is _____ has left a company as an employee.

20. Briefly describe four purposes of a business plan.

   _____
   _____
   _____
   _____

Name _____ Date _____ Class _____

Activity 38-1
Drafting Your Résumé

Preparing a résumé is an excellent way to help you consider the type of job you want. You will identify your marketable skills, outline your work history, and identify potential character references. These types of activities can help you learn more about yourself and what you can offer potential employers.

This is a two-part activity. In this part, you will prepare a draft of your résumé for a construction job. Your instructor will review your draft and then ask you to refine your résumé in Activity 38-2.

Review the sample résumé in your textbook. Then complete the following steps to prepare your draft. You may write your draft directly on a computer or begin with a pen-and-paper draft.

1. Write "Résumé—Draft 1" at the top of the page. Enter the following headings at the left margin, leaving 2–3 inches of space between each of the headings:

    A. Contact information

    B. Type of job desired

    C. Education

    D. Skills

    E. Activities

    F. References

2. Preparing the initial draft is like brainstorming. Take a few minutes to enter what first comes to mind in each of the categories. Remember, the purpose of the résumé is to outline why you are a good candidate for the job you seek.

3. Save a copy of this draft. Open a new document or use a new piece of paper to create the second draft of your résumé.

4. Review the information in the contact section.

    A. Did you provide your proper legal name? Do not use nicknames.

    B. Is your address complete? Include your zip code.

    C. Did you include all telephone numbers where you can be reached? Are these telephone numbers complete with area code?

    D. Did you include a regularly used e-mail address? If you do not have easy access to e-mail or do not check it regularly, do *not* include an e-mail address.

    E. Update the contact section to reflect your responses to these questions. Review once more to ensure that everything is correct.

Copyright by The Goodheart-Willcox Co., Inc.

5. Review the information you provided in the section on type of job desired.

   A. Is the job described an entry-level position for a person with your education, skills, and experience?

   B. Is the description specific enough for an employer to match it with the work available?

   C. Would it be better to divide the description into two or more descriptions because the qualifications required are different?

   D. Can the description be written to include more than one specific type of job because the qualifications are similar?

   E. Update this section based on your responses to these questions. If you decided that you are interested in more than one type of employment, select the one that is most likely to be available and focus your résumé on that type of employment.

6. The education section may seem difficult to write. As a current student, you may not have had the opportunity to complete many courses that relate specifically to the type of job you seek. However, there may be more you can include than you first thought.

   A. Begin by entering your current educational status.

   B. Are you specializing in a technical program related to construction? If yes, provide the name of the program. For example, "Majoring in Construction Technology."

   C. What classes have you taken that are directly related to the job you seek? Include all engineering, business, advanced math, technology, and computer classes. Include this class also.

   D. Have you participated in any classes or workshops outside of school that relate to the type of employment you seek?

   E. Update this section based on your responses to these questions.

7. Review the skills section, taking the following questions into consideration.

   A. What skills do employers expect new employees to have?

   B. Consider the various skills listed in the textbook. Of those listed, which do you have? How can you let an employer know that you possess these skills?

   C. Review the five competencies for employment listed in your textbook. What have you done to demonstrate you are sufficiently competent in each of the competencies?

   D. Focus on what you have actually done and describe skills you possess. Use active verbs.

   E. Update this section based on your responses to these questions.

Name _____

8. In the activities list, identify things you are doing or have done that are different from those listed in the education category. This includes hobbies, extracurricular activities, sports, and volunteer work. Remember that the activities should have some relationship to the type of employment you seek.

   A. Don't overlook your participation in activities involving groups of people. Such activities call attention to your ability to work with other people. Indicate the level of your participation and any recognition received. For example, "High school football—lettered three years" or "Designed and built sets for junior class play." These types of activities demonstrate the willingness to participate as well as a time commitment.

   B. Indicate any leadership roles you have had in your activities. Leadership roles demonstrate commitment, motivation, and can also act as conversation starters for an interview.

   C. Update this section based on your responses to these questions.

9. Choose your references carefully. The people you choose should know you well and understand what characteristics you possess that are likely to be important to a potential employer. A quality reference can identify those competencies for employment that you possess. Ask each person for their permission to list them as a reference. If they do not grant permission do *not* include them as a reference.

   A. A previous or current employer can be an appropriate choice. If you have not worked, choose someone who knows you are reliable, punctual, willing to work, and capable of working effectively both independently and with other people.

   B. List one reference who is familiar with any technical skills you have that apply to the type of work required by the job. This will probably be a teacher, but can be someone else such as someone with whom you have worked as a volunteer. You may also want to include an individual who is familiar with any activities and accomplishments that demonstrate your mastery of the basic skills.

   C. Provide complete and correct contact information for each reference. Indicate each references' job title and their relationship with you, so the employer will have an idea of the type of information the person is able to provide.

   D. Update this section based on your responses to these questions.

   E. Save a copy of this second draft, either on the computer or on a piece of paper.

10. If you wrote your draft on paper, create an electronic document on the computer at this time. Choose line spacing that will provide enough room for your instructor to enter suggestions and changes. Save the revised document and print a copy.

11. Submit your draft for review.

Name _____ Date _____ Class _____

Activity 38-2
Refining Your Résumé

A well-written résumé represents you in the best possible way when you are not able to represent yourself personally. In all likelihood, the person reviewing your résumé will decide whether or not to contact you based on what your résumé says and how it is written. A good résumé is:

- Brief and to the point, preferably only one page long.
- Targeted to the type of employment desired.
- Written clearly and concisely, using active verbs.
- Free of spelling or grammatical errors.
- Up-to-date.
- Organized neatly and logically organized.

Your instructor has reviewed your draft résumé and made suggestions for improvement. Include those suggestions as you complete the following steps to refine your résumé.

1. Will your résumé fit on one page? If not, look for ways to reduce the size but retain the most important information. The contact section should not have any unnecessary information.

    A. Did your instructor suggest any changes in the section on type of job desired? Does it clearly describe the type of job you want? Can the number of words be reduced? Make the needed changes.

    B. Can any education items be removed? For example, it is necessary to complete elementary school in order to be admitted to high school. Therefore, if you've listed your elementary school education, you can delete this. Employers understand that you completed elementary school in order to attend high school.

    C. Rank your skills based on their relevance to the type of job you are seeking. To what extent are the lower ranked skills necessary to learning the more important skills? If you must know how to perform one skill in order to learn another, the prerequisite skill can be removed from the list without diminishing the description of your skills.

    D. Review and rank your activities based on their relevance to the type of employment you seek. Can any activities be omitted? Rewrite the list in order of importance and eliminate those that are not essential.

    E. Does your references list include the best people to provide references for you? Three references are adequate. Are their names, titles, and contact information correct?

    F. Make changes and save the revised copy as "Draft 3."

2. If your résumé is still more than one page, formatting it differently may allow it to fit on a single page.

   A. Save your draft with a new file name, such as "Résumé—Format 1." This document can be formatted without accidentally destroying your previous draft.

   B. Formatting a document involves setting margins and tabs and defining type styles and spacing. Choose a type size and style that is easily readable and does not appear crowded. The résumé shown in the textbook uses a simple format that can be created with most word-processing software. Your document should be easy to read and well organized, and call attention to the major categories of information included.

   C. You can also save space by deleting the contact label, since your name and address are obviously contact information. Try adjusting the heading sizes, too, in order to save space.

   D. Using sentence fragments in the education, skills, and activities entries is acceptable in order to save space. Use concise, clear phrases and active verbs.

   E. Once the formatting looks reasonable, save the document. Create a new file of the same document named "Résumé—Format 2." Experiment with additional formatting ideas on this second document.

   F. If you still cannot fit your résumé on one page, move your references to the back of the page. This will free several lines of type on the front without detracting from the content. If your document is still not one page, continue experimenting with ways to reduce the information. Save your work frequently using different file names so you can return to an earlier version if you choose to do so.

   G. Once you are pleased with the formatted résumé, review the entire document one more time. Make any corrections, save the résumé, and print a copy.

3. Ask an adult to review your résumé.

4. While your résumé is being reviewed, think about additional educational goals, skills, or activities that could improve your chances of being hired for the type of job you desire. Briefly describe at least one action you could take in each category.

   A. Education

   _____
   _____

   B. Skills

   _____
   _____

   C. Activities

   _____
   _____

5. Make the changes you feel are desirable from the review in Step 3. Print a copy of the revised résumé and give it to your instructor. Make a copy for future use.

Chapter 39 Review
Construction in the Future

Name _____ Date _____ Class _____

Study Chapter 39 of the text, then answer the following questions.

1. List three prefabricated products used in the construction industry.
   _____
   _____
   _____

2. Identify two relatively new building materials or products that reduce energy consumption.
   _____
   _____

3. What can be done in building design to take advantage of solar energy other than installing solar collectors?
   _____

4. Name three sustainable sources of energy.
   _____
   _____
   _____

5. In _____, the resources used to construct and operate buildings do not negatively impact the availability of those resources for future generations.
   _____

Match each term with the correct definition.

_____ 6. A resource that can be grown for use in the future.

_____ 7. A resource than can be reused.

_____ 8. A resource that will not run out.

A. Inexhaustible resource
B. Recyclable resource
C. Renewable resource

_____ 9. Which of the following is *not* a way to improve the transportation system?
   A. Integrate highways, airports, railroads, and ships.
   B. Create water transport for passengers in the central United States.
   C. Create high-speed passenger train lines for short and medium distance trips.
   D. Create inner-city train lines and improve bus service

10. Identifying the longitude and latitude of a location directly from satellites was made possible by _____.

_____

11. Many short span bridges are now constructed using _____ beams.

_____

_____ 12. Computer software makes it possible to prepare _____.
   A. construction drawings in less time
   B. more accurate estimates, more quickly
   C. accurate summaries of the cost of each major component of the project
   D. All of the above.

Name _____ Date _____ Class _____

Activity 39-1
Construction and Your Future

In this activity, you will learn about the relationship between construction trends and employment opportunities. Changes in the construction industry directly affect construction employment. For example, the introduction of front loaders and backhoes created jobs for equipment operators but eliminated the jobs of laborers who built foundations, footings, and trenches. The introduction of plywood and other panel products simplified the installation of subflooring and roofing. Increased energy costs have created a demand for high-efficiency heating and cooling equipment. Land use restrictions in some areas have encouraged the construction of multistory residential buildings. The current trend in green construction has also had an effect on construction jobs.

Name a type of heating and cooling equipment that is likely to decline in popularity as a result of rising energy costs.

_____

Name a type of heating and cooling system likely to increase in popularity.

_____

Assume you are interested in a career in the HVAC field. Would you develop the necessary skills to install and service both systems or only one system?

_____

Assume you are interested in a plumbing-related career. Name one change in the plumbing field that is likely to affect your career.

_____

_____

The use of alternative energy sources such as wind, solar, and geothermal continues to create new construction jobs. Commercial wind generators require massive foundations that must be built and large manufactured components must be raised and assembled. Electrical wiring for each wind generator must be connected to the electric power grid.

Commercial solar installations require foundations, assembly of a prefabricated framework and solar panels, and wiring to connect the panels to the electric grid. The installation of solar collectors on individual buildings is also a growing area of employment.

The use of geothermal energy creates new ways of installing heating systems. Excavation is necessary for in-ground piping and the system must be connected to the structure.

Which of these trends is most interesting to you?

_____

Copyright by The Goodheart-Willcox Co., Inc.

Identify one job related to the trend you selected.

_____
_____

What new skill would you need to learn in order to do this job?

_____
_____
_____

New jobs created in the telecommunication field include installation of and maintenance of cell phone towers, wireless networks, home entertainment centers, and security systems. In addition to installation, buildings and other types of housing for the equipment must be constructed.

Which of these trends interests you?

_____

What new skill would you need to learn in order to do residential installation of one of these systems?

_____
_____
_____

Changes in our transportation systems include building new rail lines for travel and transition to vehicles powered by alternative forms of energy. New rail lines will create construction jobs, while alternatively powered vehicles will require construction of refueling facilities. Manufacturing facilities will also need to be modified or built.

Changes in building standards and incentive programs that encourage green construction may also impact construction jobs.

Prefabrication uses modules that are manufactured off site and delivered to a construction site where they are put in place. This creates new jobs at off-site facilities and creates new jobs on site for workers who place and attach the modules to the remainder of the building.

Which of these trends interests you?

_____

As a carpenter, name one way your job would change if you were working on one of these projects rather than traditional residential construction.

_____
_____

An important lesson to learn from this activity is that change will happen and the rate of change will continue to accelerate. To take advantage of future employment opportunities, keep an eye on future trends and continually upgrade your skills.

Construction Company Simulation

The Construction Company Simulation section provides realistic experiences in reading drawings and specifications, estimating and bidding, contracting, managing employees, accounting for labor and materials costs, and building activities. The focus of these activities is to provide you with opportunities to practice a variety of construction and management skills. You will learn how to evaluate your own skills, to work cooperatively with your classmates, and to help one another by providing constructive suggestions.

The following drawings, specifications, worksheets, certificates, spreadsheets, and other forms are for use with the Construction Company Simulation activities. Your instructor will provide specific instructions on their use.

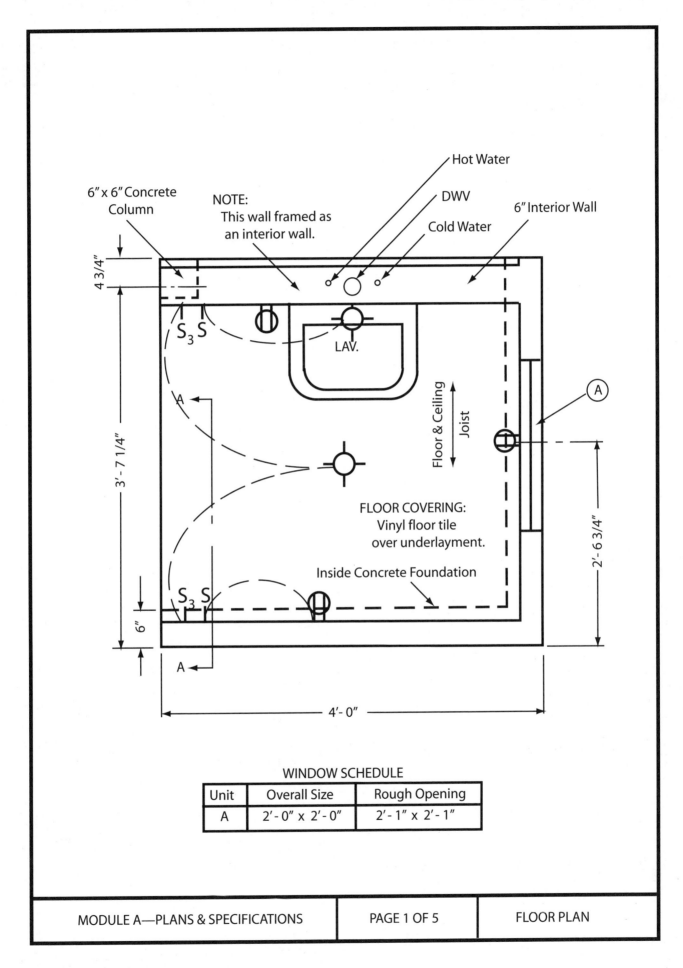

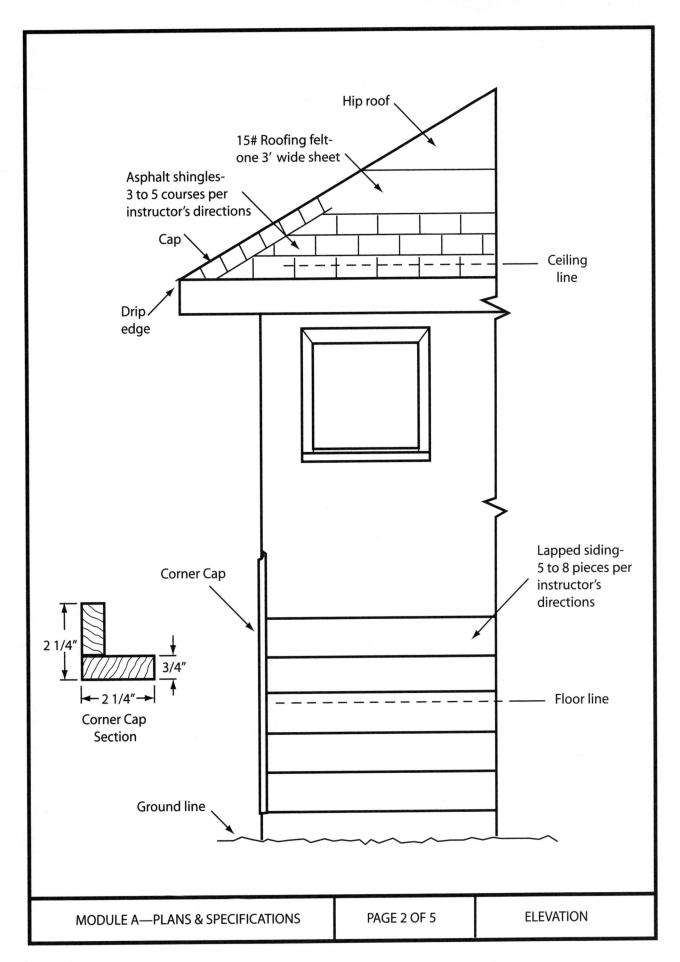

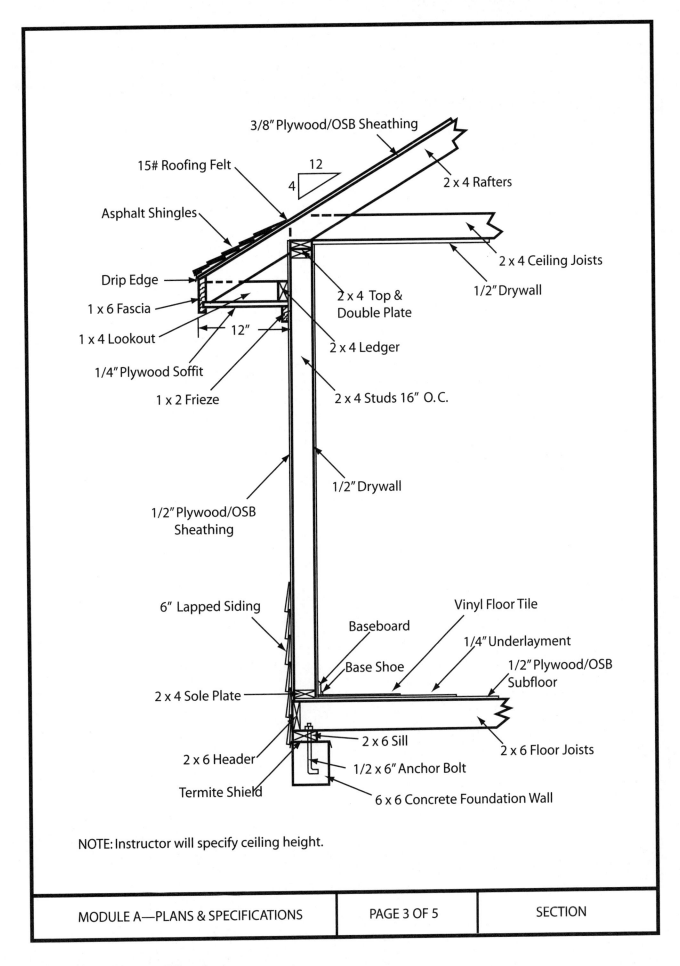

NOTES:
1. 1/2" CPVC hot water pipe & fittings.
2. 1/2" copper cold water pipe & fittings.
3. 1 1/2" PVC DWV pipe & fittings.
4. Water supply stub-outs 12" above the floor & 4" either side of DWV pipe center.
5. DWV stub-out 14" above floor.
6. DWV pipe extends 12" above roof.
7. Boot to be installed around DWV pipe above roof.
8. Top of sink bracket 32" above the floor.

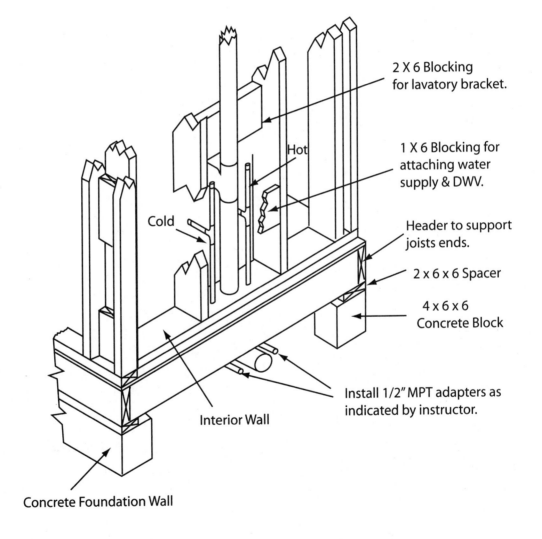

1. FOUNDATION
   A. Forms—Build new forms or use existing forms as specified by instructor.
   B. Shop floor protection—3/4" × 6' × 6' plywood covered with 6' × 6' polyethylene sheet.
   C. Concrete—6 bag mix.
   D. Anchor bolts—4-6". Consult floor framing contractor for location.

2. FLOOR FRAME
   A. Lumber—Utility grade 2 × 6s.
   B. Subfloor—3/8" plywood or OSB.
   C. Fasteners—16d box nails for 2xs, 6d box or 1 1/4" screws for subfloor.

3. EXTERIOR WALLS
   A. Lumber—Exterior walls, 2 × 4s. Interior walls, utility grade 2 × 6s.
   B. Sheathing—3/8" plywood or OSB.
   C. Window—Rough opening 10" below ceiling joists.
   D. Fasteners—16d box nails for 2 × 4s, 6d box nails or 1 1/4" screws for sheathing.

4. ROOF
   A. Lumber—Utility grade 2 × 4s.
   B. Sheathing—3/8" plywood or OSB.
   C. Drip edge—T-shaped galvanized steel or painted aluminum.
   D. 15# roofing felt—One 3' wide strip beginning at lower edge of roof.
   E. Shingles—Composition, 3-tab. Three to five courses as specified by instructor.

5. EXTERIOR FINISH
   A. Siding—8" lapped siding. Five to ten courses as specified by instructor.
   B. Soffit—1/4" plywood.
   C. Fasteners—6d box nails for siding. 8d finish nails for corner cap and fascia. 4d nails for soffit.

6. INTERIOR FINISH
   A. Premixed joint cement.
   B. Trim—Baseboard, 2 1/4" primed. Window casing, 2 1/4" primed.
   C. Underlayment—1/4" plywood.
   D. Floor tile—12" × 12" vinyl.
   E. Fasteners—1 1/8" drywall screws. 6d finish nails for baseboard. 4d and 6d finish nails for casing.

7. PLUMBING
   A. 1 1/2" PVC pipe and fittings.
   B. Cold water supply—1/2" copper pipe and fittings.
   C. Hot water supply—1/2" CPVC pipe and fittings.
   D. Lavatory—As specified by instructor.
   E. Angle stops—FPT adapter at stub-outs.

8. ELECTRICAL
   A. Wiring—#14-2 w/ground.
   B. Boxes—Single gang, double gang, and octagonal.
   C. Receptacles—Duplex.
   D. Switches—Single pole and three-way.
   E. Light above lavatory—Centered in wall above lavatory, 12" below ceiling.
   F. Ceiling light—Centered in ceiling.
   G. Fasteners—Wire staples and nails or screws to attach boxes.

MODULE A—PLANS & SPECIFICATIONS	PAGE 5 OF 5	SPECIFICATIONS

Copyright by The Goodheart-Willcox Co., Inc.

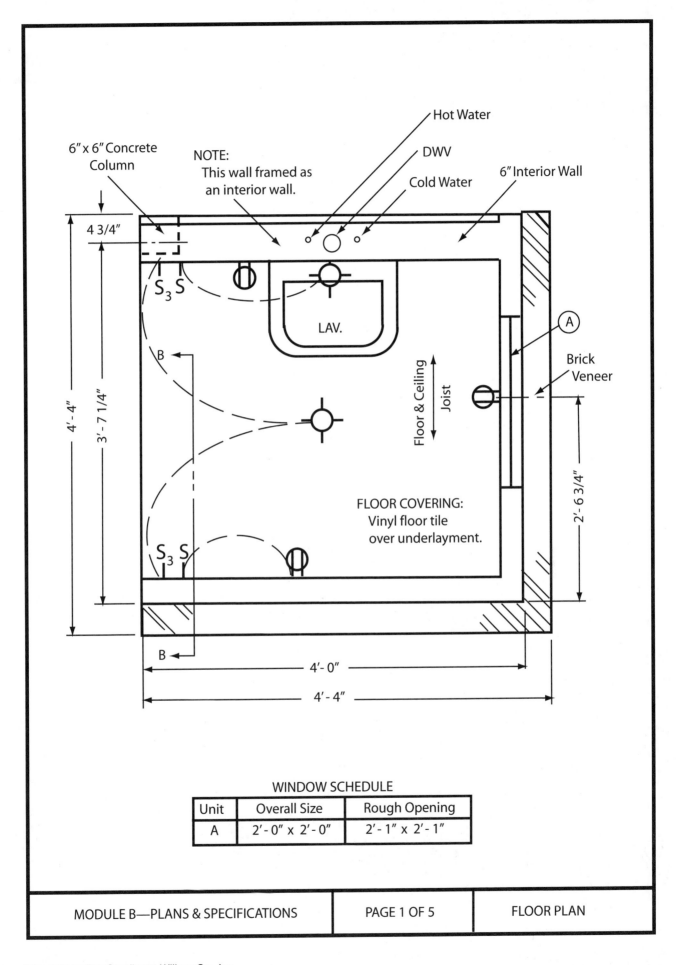

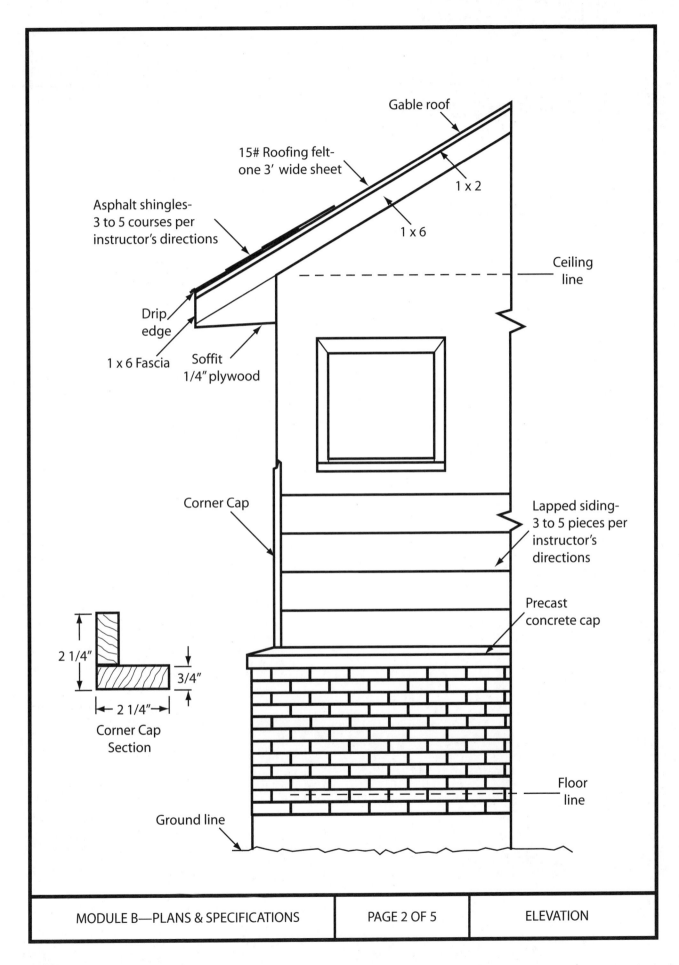

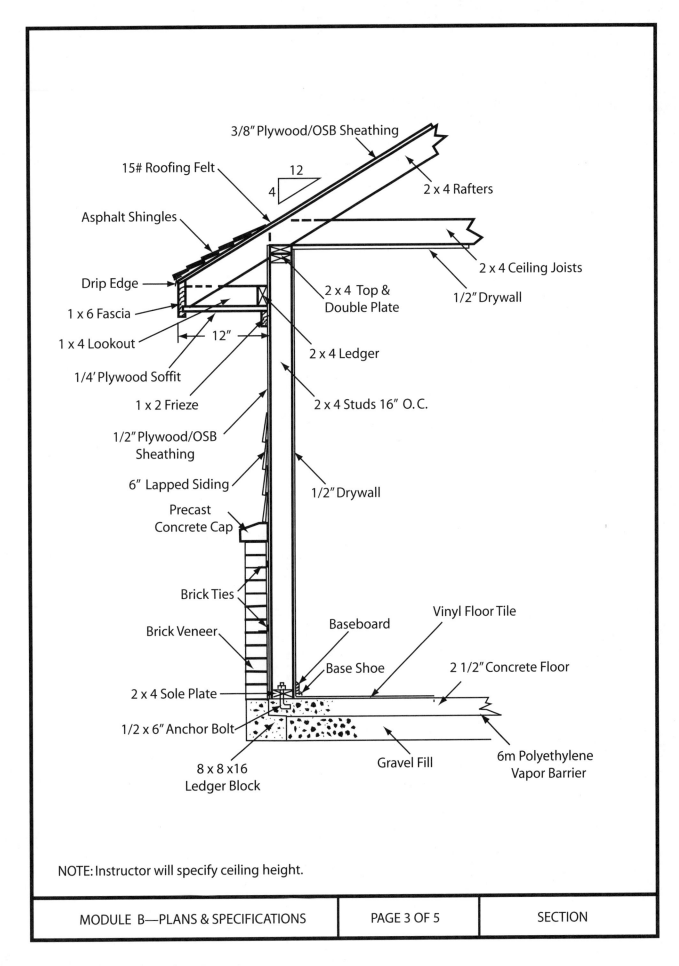

NOTES:
1. 1/2" CPVC hot water pipe & fittings.
2. 1/2" copper cold water pipe & fittings.
3. 1 1/2" PVC DWV pipe & fittings.
4. Water supply stub-outs 12" above the floor & 4" either side of DWV pipe center.
5. DWV stub-out 14" above floor.
6. DWV pipe extends 12" above roof.
7. Boot to be installed around DWV pipe above roof.
8. Top of sink bracket 32" above the floor.

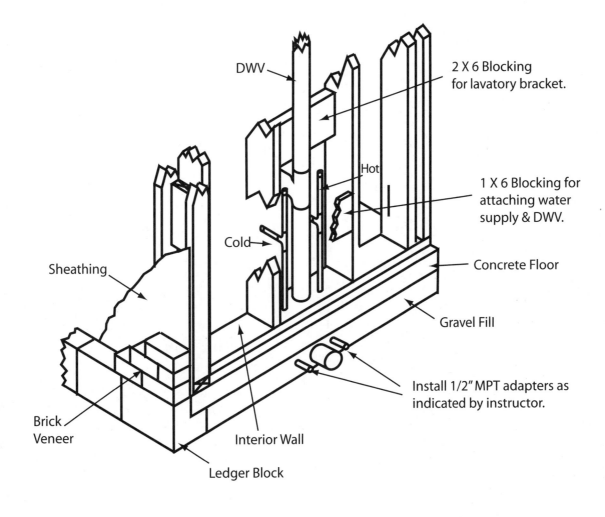

| MODULE B—PLANS & SPECIFICATIONS | PAGE 4 OF 5 | PLUMBING PLAN |

1. FOUNDATION
   A. Ledger block—8″ × 8″ × 16″.
   B. Shop floor protection—3/4″ × 6′ × 6′ plywood covered with 6′ × 6′ polyethylene sheet.
   C. Mortar—1 part masonry cement, 1 1/2 parts lime, 5 parts sand, plus water. This makes a weak form of mortar that is easily removed during salvage.
   D. Anchor bolts—4-6″. Consult floor framing contractor for location.

2. FLOOR FRAME
   A. 3/4″ gravel fill.
   B. Concrete—Use formulas from concrete activity to calculate amount of Portland cement, sand, gravel, and water needed.
   C. Finish—Smooth troweled.

3. EXTERIOR WALLS
   A. Lumber—Exterior walls, 2 × 4s. Interior walls, utility grade 2 × 6s.
   B. Sheathing—3/8″ plywood or OSB.
   C. Window—Rough opening 10″ below ceiling joists.
   D. Fasteners—16d box nails for 2 × 4s, 6d box nails or 1 1/4″ screws for sheathing.

4. ROOF
   A. Lumber—utility grade 2 × 4s.
   B. Sheathing—3/8″ plywood or OSB.
   C. Drip edge—T-shaped galvanized steel or painted aluminum.
   D. 15# roofing felt—One 3′ wide strip beginning at lower edge of roof.
   E. Shingles—Composition, 3-tab. Three to five courses as specified by instructor.

5. EXTERIOR FINISH
   A. Brick veneer—Common brick. Three brick ties every fourth course. For mortar, see Foundation above.
   B. Siding—6″ lapped siding. Four to six courses as specified by instructor.
   C. Soffit—1/4″ plywood.
   D. Fasteners—6d box nails for siding. 8d finish nails for corner cap and fascia. 4d nails for soffit.

6. INTERIOR FINISH
   A. Premixed joint cement.
   B. Trim—Baseboard, 2 1/4″ primed. Window casing, 2 1/4″ primed.
   C. Underlayment—1/4″ plywood.
   D. Floor tile—12″ × 12″ vinyl.
   E. Fasteners—1 1/8″ drywall screws. 6d finish nails for baseboard. 4d and 6d finish nails for casing.

7. PLUMBING
   A. 1 1/2″ PVC pipe and fittings.
   B. Cold water supply—1/2″ copper pipe and fittings.
   C. Hot water supply—1/2″ CPVC pipe and fittings.
   D. Lavatory—As specified by instructor.
   E. Angle stops—FPT adapter at stub-outs.

8. ELECTRICAL
   A. Wiring—14-2 w/ground.
   B. Boxes—Single gang, double gang, and octagonal.
   C. Receptacles—Duplex.
   D. Switches—Single pole and three-way.
   E. Light above lavatory—Centered in wall above lavatory, 12″ below ceiling.
   F. Ceiling light—Centered in ceiling.
   G. Fasteners—Wire staples and nails or screws to attach boxes.

MODULE B—PLANS & SPECIFICATIONS	PAGE 5 OF 5	SPECIFICATIONS

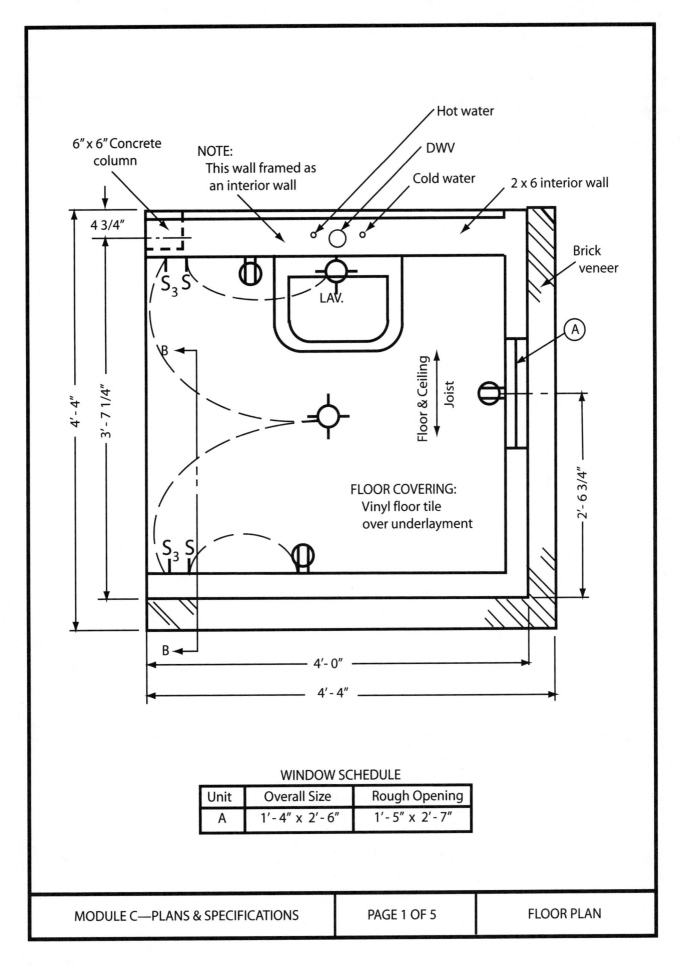

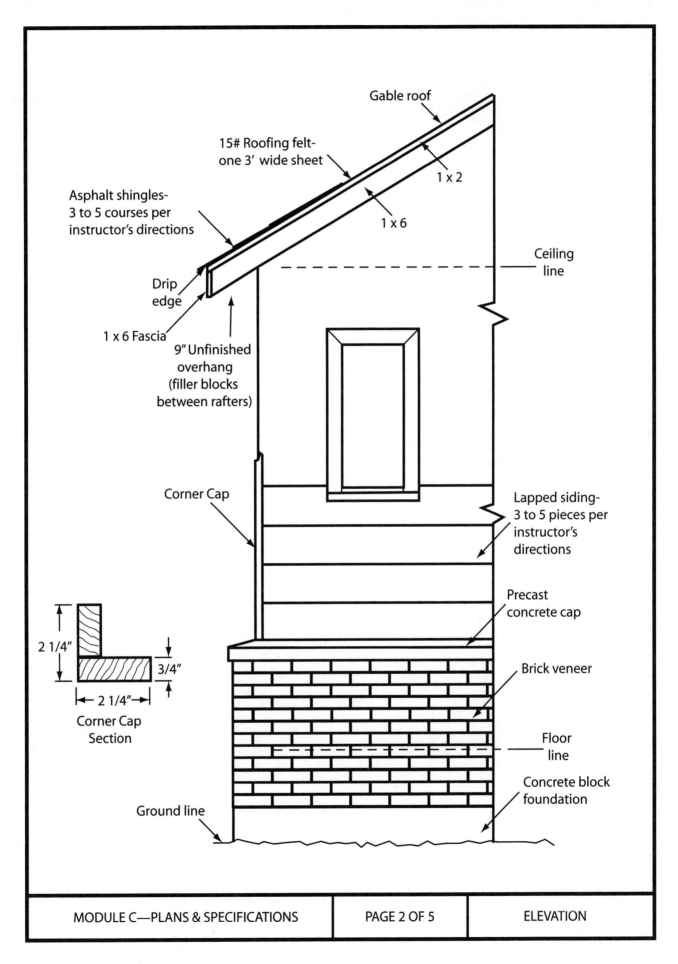

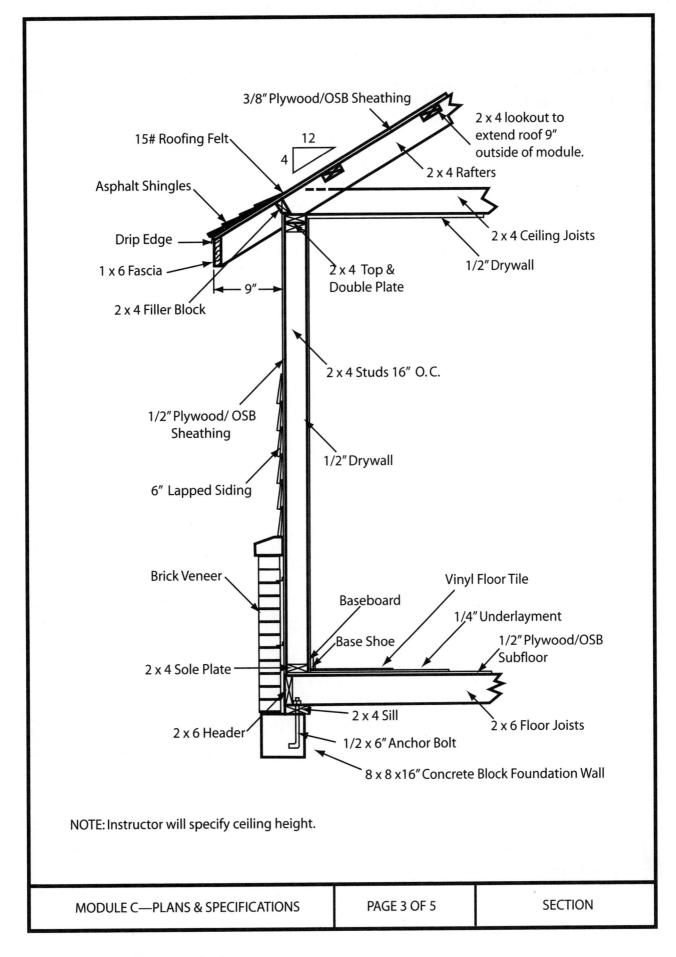

NOTES:
1. 1/2" CPVC hot water pipe & fittings.
2. 1/2" copper cold water pipe & fittings.
3. 1 1/2" PVC DWV pipe & fittings.
4. Water supply stub-outs 12" above the floor & 4" either side of DWV pipe center.
5. DWV stub-out 14" above floor.
6. DWV pipe extends 12" above roof.
7. Boot to be installed around DWV pipe above roof.
8. Top of sink bracket 32" above the floor.

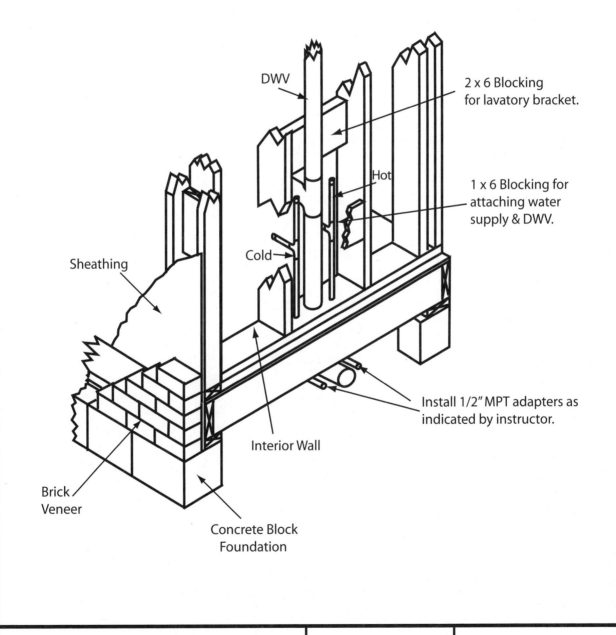

MODULE C—PLANS & SPECIFICATIONS | PAGE 4 OF 5 | PLUMBING PLAN

1. FOUNDATION
   A. Wall—8 × 8 × 16 concrete block.
   B. Shop floor protection—3/4" × 6' × 6' plywood covered with 6' × 6' polyethylene sheet.
   C. Mortar—1 part masonry cement, 1 1/2 parts lime, 5 parts sand, plus water. This makes a weak form of mortar that is easily removed during salvage.
   D. Anchor bolts—4-6". Consult floor framing contractor for location.

2. FLOOR FRAME
   A. Lumber—Utility grade 2 × 4s and 2 × 6s.
   B. Subfloor—3/8" plywood or OSB.
   C. Fasteners—16d box nails for 2xs, 6d box or 1 1/4" screws for subfloor.

3. EXTERIOR WALLS
   A. Lumber—Exterior walls, 2 × 4s. Interior walls, utility grade 2 × 6s.
   B. Sheathing—3/8" plywood or OSB.
   C. Window—Rough opening 10" below ceiling joists.
   D. Fasteners—16d box nails for 2 × 4s, 6d box nails or 1 1/4" screws for sheathing.
   E. Blocking—2xs between rafters, flush with outside of frame.

4. ROOF
   A. Lumber—Utility grade 2 × 4s.
   B. Sheathing—3/8" plywood or OSB.
   C. Drip edge—T-shaped galvanized steel or painted aluminum.
   D. 15# roofing felt—One 3' wide strip beginning at lower edge of roof.
   E. Shingles—Composition, 3-tab. Three to five courses as specified by instructor.

5. EXTERIOR FINISH
   A. Siding—8" lapped siding. Five to ten courses as specified by instructor.
   B. Fasteners—6d box nails for siding. 8d finish nails for corner cap and fascia. 4d nails for soffit.

6. INTERIOR FINISH
   A. Premixed joint cement.
   B. Trim—Baseboard, 2 1/4" primed. Window casing, 2 1/4" primed.
   C. Underlayment—1/4" plywood.
   D. Floor tile—12" × 12" vinyl.
   E. Fasteners—1 1/8" drywall screws. 6d finish nails for baseboard. 4d and 6d finish nails for casing.

7. PLUMBING
   A. 1 1/2" PVC pipe and fittings.
   B. Cold water supply—1/2" copper pipe and fittings.
   C. Hot water supply—1/2" CPVC pipe and fittings.
   D. Lavatory—As specified by instructor.
   E. Angle stops—FPT adapter at stub-outs.

8. ELECTRICAL
   A. Wiring—#14-2 w/ground.
   B. Boxes—Single gang, double gang, and octagonal.
   C. Receptacles—Duplex.
   D. Switches—Single pole and three-way.
   E. Light above lavatory—Centered in wall above lavatory, 12" below ceiling.
   F. Ceiling light—Centered in ceiling.
   G. Fasteners—Wire staples and nails or screws to attach boxes.

MODULE C—PLANS & SPECIFICATIONS	PAGE 5 OF 5	SPECIFICATIONS

Copyright by The Goodheart-Willcox Co., Inc.

Construction Company Simulation 389

Estimating Work Sheet

Name _____   Student I.D. # _____

Materials

Job Cat.	No. Pcs.	Size (T × W × L)	Qty.	Description	Unit Cost	Material Total	No. Emp.	Hr./ Emp.	Total Hours	Cost/ Hour	Labor Total	Job Total

Management

Est. Time	Cost/ Hour	Mgn. Cost
		$

Summary

Management $
Materials Cost $
Labor $

Estimated Total Co: $

Copyright by The Goodheart-Willcox Co., Inc.

Estimating Work Sheet

Name _____ Student I.D. # _____

Materials

Job Cat.	No. Pcs.	Size (T × W × L)	Qty.	Description	Unit Cost	Material Total	No. Emp.	Hr./ Emp.	Total Hours	Cost/ Hour	Labor Total	Job Total

Management

Est. Time	Cost/ Hour	Mgn. Cost
	$	

Summary

Management	$
Materials Cost	$
Labor	$
Estimated Total Co:	$

Job Categories and Standard Data

Job Categories	Job Category Number	Time per Module (minutes)		
		A	B	C
General contractor—management	1.1	200	200	200
General contractor—labor	1.2	300	350	350
Concrete—management	2.1	75	65	N/A
Concrete—labor	2.2	750	565	N/A
Masonry—management	3.1	N/A	85	85
Masonry—labor	3.2	N/A	650	650
Floor framing—management	4.1	45	N/A	45
Floor framing—labor	4.2	385	N/A	385
Wall framing—management	5.1	120	120	120
Wall framing—labor	5.2	950	950	950
Roof and ceiling framing—management	6.1	150	65	65
Roof and ceiling framing—labor	6.2	1000	650	650
Plumbing—management	7.1	65	85	65
Plumbing—labor	7.2	280	350	280
Electrical—management	8.1	75	75	75
Electrical—labor	8.2	330	330	330
Salvage—management	9.1	60	60	60
Salvage—labor	9.2	330	330	330
Special Job Categories				
Student trainer	25			
Clerk	26			
Accounting	27			
Records	28			
Clean-up	29			
Instructor assigned	30			

Materials Price List

Item Number	Material
Concrete and Masonry	
104	Masonry sand—$4.86/cubic foot
108	Portland cement—$8.45/cubic foot
112	Masonry cement—$6.17/cubic foot
116	Hydrated lime—$5.69/cubic foot
120	Gravel 3/4"—$3.75/cubic foot
124	Rebar 1/2"—$0.45/cubic foot
128	8 × 8 × 8, 1-core concrete block—$0.95 each
132	8 × 8 × 16, 3-core concrete block—$1.43 each
136	8 × 8 × 16 ledger block—$2.36 each
140	Face brick—$0.38 each
144	Precast concrete cap—square ends, $2.68 each, 45° one end, $2.76 each
148	Fasteners and supplies—$1.00/concrete or masonry contract
Framing Materials	
204	2 × 4 × 6'—$2.61 each
208	2 × 4 × 8'—$2.34 each
212	2 × 4 × 10'—$3.78 each
216	2 × 4 × 14'—$4.09 each
220	2 × 6 × 8'—$3.48 each
224	2 × 6 × 10'—$4.32 each
228	3/8" × 4' × 8' OSB or plywood sheathing or subflooring—$12.35/sheet
232	Fasteners and supplies—$5.15/wall framing contract, $2.80/floor or roof framing contract
Roofing	
304	15# roofing felt—$0.13/lineal foot
308	Drip edge—$0.42/lineal foot
312	Asphalt shingles—$0.80/square foot of coverage
316	Fasteners and supplies—$1.45/general contract
Exterior Finish	
404	Window 24" × 24"—$15.75 each
408	Window 16" × 30"—$18.95 each
412	1 × 6 × 10' B or Better softwood lumber—$8.25 each
416	1 × 6 × 12' beveled siding—$6.92 each
420	1 × 8 × 12' beveled siding—$7.49 each
424	Fasteners and supplies—$1.65/general contract
Plumbing	
504	1/2" copper tube—$1.25/foot
508	1/2" × 1/2" × 1/2" copper tee—$0.98 each
512	1/2" × 1/2" copper 90° ell—$0.58 each

(Continued)

Materials Price List (Continued)

Item Number	Material
	Plumbing
516	1/2″ copper coupling—$0.76 each
520	1/2″ sweat × 1/2 MPT copper adapter—$2.99 each
524	1/2″ pipe straps—$0.11 each
528	1/2″ copper cap—$0.35 each
532	1 1/2″ PVC pipe—$0.44/foot
536	1 1/2″ × 1 1/2″ × 1 1/2″ PVC sanitary tee—$2.53 each
540	1 1/2″ × 1 1/2″ PVC 90° ell—$1.25 each
544	1 1/2″ PVC coupling—$0.86 each
548	1 1/2″ PVC cap—$1.00 each
552	1 1/2″ pipe straps—$0.35 each
556	Lavatory wall hung with bracket—$32.00 each
560	Faucet with drain assembly—$23.68 each
564	1/2″ angle valve—$6.28 each
568	1 3/8″ plastic P-trap—$3.44 each
572	1 3/8″ × 1 1/2″ adapter—$2.47 each
576	Fasteners and supplies—$2.25/plumbing contract
	Electrical
604	Outlet box, plastic—$0.49 each
608	2-gang outlet box, plastic—$0.98 each
612	Octagonal box, 4″ plastic—$0.98 each
616	Ceiling bracket, fiberglass—$3.67 each
620	Duplex outlet—$0.68 each
624	GFCI duplex outlet—$2.86 each
628	#14 2-wires w/ground—$0.39/foot
632	#14 3-wires w/ground—$0.61/foot
636	Bell wire 18-2—$0.25/foot
640	Light fixture—$2.58 each
644	Single-pole light switch—$0.49 each
648	3-way light switch—$1.64 each
652	Doorbell—$19.97 each
656	Doorbell switch—$4.34 each
660	Doorbell transformer—$13.92 each
664	Fasteners and supplies—$1.80/electrical contract
	Interior Finish
704	1/2″ × 4 × 8 drywall—$7.40/sheet
708	1/4″ × 4 × 8 underlayment—$8.89 each
712	2 1/4″ baseboard—$0.39/foot
716	2 1/4″ window casing—$0.41/foot
720	Fasteners and supplies—$2.65/general contract

Construction and Building Technology
Bid Blank

Date _____

Having carefully examined the drawings and specifications entitled _____, as well as the site and conditions affecting the work, the undersigned proposes to furnish all materials and labor required to complete the _____ phase of the project in accordance with said documents for the sum of $ _____. If I am notified of acceptance of this proposal within two (2) class periods, I agree to execute a contract for the above stated work. "Compensation" is to be in the form of credit toward may grade for this class.

Submitted by,

_____

Contractor I.D. #

_____

Construction and Building Technology
Bid Blank

Date _____

Having carefully examined the drawings and specifications entitled _____, as well as the site and conditions affecting the work, the undersigned proposes to furnish all materials and labor required to complete the _____ phase of the project in accordance with said documents for the sum of $ _____. If I am notified of acceptance of this proposal within two (2) class periods, I agree to execute a contract for the above stated work. "Compensation" is to be in the form of credit toward may grade for this class.

Submitted by,

_____

Contractor I.D. #

_____

Copyright by The Goodheart-Willcox Co., Inc.

Construction and Building Technology
Bid Blank

Date _____

Having carefully examined the drawings and specifications entitled _____, as well as the site and conditions affecting the work, the undersigned proposes to furnish all materials and labor required to complete the _____ phase of the project in accordance with said documents for the sum of $ _____. If I am notified of acceptance of this proposal within two (2) class periods, I agree to execute a contract for the above stated work. "Compensation" is to be in the form of credit toward may grade for this class.

Submitted by,

_____

Contractor I.D. #

_____

Construction and Building Technology
Bid Blank

Date _____

Having carefully examined the drawings and specifications entitled _____, as well as the site and conditions affecting the work, the undersigned proposes to furnish all materials and labor required to complete the _____ phase of the project in accordance with said documents for the sum of $ _____. If I am notified of acceptance of this proposal within two (2) class periods, I agree to execute a contract for the above stated work. "Compensation" is to be in the form of credit toward may grade for this class.

Submitted by,

_____

Contractor I.D. #

_____

Copyright by The Goodheart-Willcox Co., Inc.

Construction and Building Technology

Contract

This agreement made on the _____ day of _____ in the year _____ by and between _____, hereinafter called the Instructor and _____, hereinafter called the Contractor. Witness, that whereas the Instructor intends to erect _____, hereinafter called the Project, now, therefore, the Instructor and the Contractor for the considerations hereinafter named, agree as follows:

1. The Contractor agrees to furnish all materials and labor and complete the _____ phase of the above named project on or before _____ periods of class-work time have been completed. Failure to complete the work on schedule will result in a penalty of $_____ per work day that the work remains incomplete.

2. The Instructor agrees to "pay" the Contractor for such services a fee, in class currency, equivalent to $_____. Class currency will apply to the Contractor's grade for this class.

In witness whereof the parties hereto have made and executed this agreement, the day and year first mentioned above.

_____   _____
Instructor   Contractor

Contractor's I.D. # _____

Construction Company Simulation 403

Name _____ Contract Type _____ Student I.D. # _____

Bar Chart Scheduling Form

Job	Time	Days																	
		1	2	3	4	5	6	7	8	9	10	11	12	13	14	15	16	17	18
	Est																		
	Act																		
	Est																		
	Act																		
	Est																		
	Act																		
	Est																		
	Act																		
	Est																		
	Act																		
	Est																		
	Act																		
	Est																		
	Act																		
	Est																		
	Act																		

Est = Estimated Time
Act = Actual Time

Copyright by The Goodheart-Willcox Co., Inc.

You will need one copy of this time card for each day you work. A sample is shown here. Your instructor will provide additional copies for your use.

Daily Time Card						
Name				**StuID#**		
Date				**Class Period**		
ContID#	JobCat	Mod	Time	EmpInit	EntITS	EntCSF
Specials						
130	26	T		N/A		
130	27	K		N/A		
130	28	R		N/A		
130	29	U		N/A		
130	30	X				

StuID# = Student ID# ContID# = Contractor ID# JobCat = Job Category
EmpInit = Employer's initials EntITS = Entered Individual Time Summary
EntCSF = Entered Contract Summary Form Mod = Module

Construction and Building Technology

Building Permit Application

Contractor _____ Date _____

    The above named contractor/subcontractor is licensed as a _____ contractor.

    The above named contractor hereby applies for a permit to do the _____ phase for _____. The following documents are attached and become part of this application:

1. Working drawings
2. Estimating worksheet
3. Contract
4. Schedule

    In making this application, the contractor agrees to perform high-quality work that meets or exceeds the standards established by the textbook for this class, previous laboratory activities, and class instruction.

    It is further understood that the contractor will notify the Building Inspector at appropriate times so inspections of work in progress can be made.

_____
Signature of Contractor
Contractor ID# _____

Date application approved _____ By _____

Purchase Order

Name _____ Contractor ID# _____

Date submitted _____ Date needed _____

Structure _____ Purchase order number (PO#) _____

Quantity	Item #	Size	Description	Unit Cost	Total
				Grand Total	

Filled by _____ I.D.# _____ Received by _____ I.D.# _____

Calculations checked by _____ I.D.# _____ Data entered into Contract Summary and checked by StuID# _____

Copyright by The Goodheart-Willcox Co., Inc.

Purchase Order

Name _____ Contractor ID# _____

Date submitted _____ Date needed _____

Structure _____ Purchase order number (PO#) _____

Quantity	Item #	Size	Description	Unit Cost	Total
				Grand Total	

Filled by _____ I.D.# _____ Received by _____ I.D.# _____

Calculations checked by _____ I.D.# _____ Data entered into Contract Summary and checked by StuID# _____

Copyright by The Goodheart-Willcox Co., Inc.

Purchase Order

Name _____ Contractor ID# _____

Date submitted _____ Date needed _____

Structure _____ Purchase order number (PO#) _____

Quantity	Item #	Size	Description	Unit Cost	Total
				Grand Total	

Filled by _____ I.D.# _____ Received by _____ I.D.# _____

Calculations checked by _____ I.D.# _____ Data entered into Contract Summary and checked by StuID# _____

Copyright by The Goodheart-Willcox Co., Inc.

Purchase Order

Name _____ Contractor ID# _____

Date submitted _____ Date needed _____

Structure _____ Purchase order number (PO#) _____

Quantity	Item #	Size	Description	Unit Cost	Total
				Grand Total	

Filled by _____ I.D.# _____ Received by _____ I.D.# _____

Calculations checked by _____ I.D.# _____ Data entered into Contract Summary and checked by StuID# _____

Name _____ Module _____ Contractor ID# _____

Store Accounting Form

1. Enter your information in the blanks above and submit this form to your instructor to become part of the Store Accounting Ledger.

2. The store clerk records each purchase order (P.O.) you place by filling in the first blank row of the table. The total-to-date cell is computed by adding the total for the current P.O. to the previous total-to-date cell. The clerk will note any questions regarding th P.O. in the notes cell.

Delivery Date	P.O.#	Total this P.O.	Total-to-Date	Clerk's I.D.#	Notes

Copyright by The Goodheart-Willcox Co., Inc.

Accounting System

Contractors pay employees for their work and suppliers for the materials they purchased to complete each contract. The accounting system developed for this class follows a procedure for accomplishing these tasks. The system also provides your instructor with information that will help determine your grade.

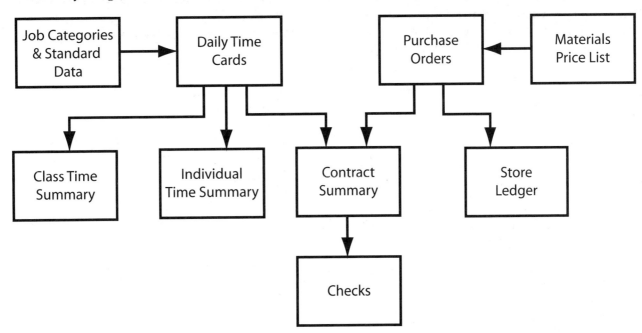

Begin by recording the time you work in each class on a daily time card (DTC). Use the student identification number *(StuID#)* assigned by your instructor. Record the contractor identification number *(ContID#)*, job category *(JobCat)*, module identification letter (A, B or C), and time spent on this job. Get the structure identification letter and contractor identification number from the employing contractor. Review the Job Categories and Standard Data sheet for job numbers. Numbers 1.1–9.2 are for work done on the modules. The Specials category list other tasks completed during lab work time. These jobs use numbers 25–30 and are listed on all DTC for your convenience.

The contractor you are working for will initial your DTC to indicate they approve the time. They will add information from your DTC to their Contract Summary. Your instructor collects the DTCs at the end of each class period.

Contractors use the purchase orders to order materials needed for jobs. Purchase order numbers *(PO#)* are assigned sequentially, beginning with 1. Item numbers *(Item #)* are found on the Materials Price List. The Materials Price List also lists available sizes and units of each material and a description. The contractor calculates the total for each item purchased and the grand total for all items on the purchase order. Order only those items needed for one class period to avoid having too many materials cluttering the job site.

The store clerk fills the purchase order, initials the *Filled By* blank, and enters their own student identification number. The purchasing contractor checks the delivery and if it is correct initials the *Received By* blank. The store clerk checks the cost calculations and if they are correct, initials the *Calculations Checked By* blank and enters their own student identification number. Any errors are worked out between the contractor and store clerk. If an agreement cannot be reached, a note is entered on the purchase order detailing the problem. Your instructor will follow up as needed.

If all calculations are correct, the store clerk enters summary information into the Store Accounting Form. The purchase order is returned to the contractor and the contractor enters the necessary information into the *Materials Purchased* box of the Contract Summary Form.

During the next class period, the accountants enter the data from the DTCs and purchase orders into a computerized database, producing individual time summaries and a class time summary. A copy of the class time summary will be posted periodically so students can check the data for accuracy. You are responsible for checking the accuracy of your time.

Remember to enter the time you worked in each job category and a total time for each day at the bottom of the form. At the end of the building activity, the job category work totals are also entered. These data are used to award bonus points for variety of work.

When work is completed on the contract, contractors write checks to each employee, to the store for materials, and to themselves for management time.

Entering Data in Spreadsheets

Excel spreadsheets are used to compile and summarize time, income, and cost data about the work students do during module construction. Your instructor uses the records to monitor your progress and make decisions about the grade you will receive, based on the objectives of these activities.

During each class period, two students are employed as accountants. They enter Daily Time Card data from the previous class into the ITS spreadsheet. The accountants work as a team to complete the entries.

The Individual Time Summary (ITS) summarizes the time worked at a particular job by a specific student. The data for the ITS is taken from the student's Daily Time Card (DTC). Work is categorized by job category (*Job Cat*). The spreadsheet automatically computes a subtotal for work on the modules (*STofM*), a subtotal for special work (*STofS*) and a daily total. The daily total allows the instructor to monitor each student's level of participation. An example of a completed ITS is shown here.

Individual Time Summary

Name: Joe   I.D.: 101   Class: 2nd

Job Cat	Mod	Emp #	9/15 Time	9/16 Time	9/18 Time	9/19 Time	9/22 Time	9/23 Time	9/24 Time	9/25 Time	9/29 Time	9/30 Time	10/2 Time	10/4 Time	10/5 Time	10/7 Time	10/10 Time	10/11 Time	10/13 Time	Total
1	B	1			25															25
2	C	5	20																	20
3	A			45																45
4	C																			0
5	A					30														30
6	A	6					30													30
7	C		30					20												50
8	A	23				10														10
9	C								30											30
10	A				10		15	35												60
11	C									15										15
12	C						20													20
13	A						10													10
14	C												40							40
15	A									40						25				65
16	C									10										10
17	A										30					20				50
18	A											5								5
19	C														15					15
20	C												10							10
21	A														15	15				30
22	A												25			20				45
23	A														30			25		55
24	C													35						35
25	C																	30		30
Special		STotM	50	55	65	65	65	30	0	65	30	0	55	60	60	60	20	55	0	735
26	T	27									40	60					15	50		165
27	K	28					30	60								25				115
28	T	29	10	10	10	5	5	10	10	10	10	10	10	10	5	10	5	10		150
29	U	30	15	10	10	15	10	5	5	5	5	5	10	5	10	5	5	5	5	130
30	X																			
		STotS	25	20	20	20	15	45	75	15	55	75	15	20	15	15	40	25	65	
		Total	75	75	85	85	80	75	75	80	85	75	70	80	75	75	60	80	65	1295

Time Ent. By

Job Cat = Job category   I.D. # = Student I.D. number   R = Records
Mod = Module   T = Student Trainer   U = Clean-up
Emp # = Employer I.D. num   K = Accounting   X = Instructor assigned

Follow this procedure to complete the spreadsheets.

1. Sort DTCs by the student identification numbers (*StuID#*) shown in the upper right-hand corner of the card.

2. Open the Excel spreadsheet for your class, find the correct *StuID#* along the lower edge of the Excel window, and click on that number. Check the DTC information and the Excel heading to ensure you have the correct spreadsheet.

3. In the Date row across the top of the ITS spreadsheet, find the first blank cell and enter today's date.

4. Enter the time data from each entry on the DTC in the cell for this date that corresponds to the correct *Job Cat* number. If more than one time is given for a job category, add the times and enter the total. Check your work carefully.

5. Enter your identification number in the *TimeEntBy* cell. Save the file. Proceed to the next DTC and ITS.

6. When you have entered all the data from the DTCs, return them to your instructor.

Individual Time Summary

Name _____ I.D. _____ Class _____

Job Cat	Date Mod	Emp #	Time	Time	Time	Time	Time	Time	Time	Time	Time	Time	Time	Time	Time	Time	Time	Time	Time	Total
1																				0
2																				0
3																				0
4																				0
5																				0
6																				0
7																				0
8																				0
9																				0
10																				0
11																				0
12																				0
13																				0
14																				0
15																				0
16																				0
17																				0
18																				0
19																				0
20																				0
21																				0
22																				0
23																				0
24																				0
25																				0
Special	STotM	0	0	0	0	0	0	0	0	0	0	0	0	0	0	0	0	0	0	0
26	T	27																		
27	K	28																		
28	T	29																		
29	U	30																		
30	X																			
	STotS	0	0	0	0	0	0	0	0	0	0	0	0	0	0	0	0	0	0	0
	Total	0	0	0	0	0	0	0	0	0	0	0	0	0	0	0	0	0	0	0

Time Ent. By _____

Job Cat = Job category   I.D. # = Student I.D. number   R = Records   U = Clean-up

Mod = Module   T = Student Trainer   X = Instructor assigned

Emp # = Employer I.D. num   K = Accounting

Copyright by The Goodheart-Willcox Co., Inc.

Construction Company Simulation

Check 1

Contractor I.D. # _____

Check No. _____

Date _____, 20 _____

Paid to the Order of _____ $ _____

_____ points credit for construction class.

Bank
TechEd   Total minutes worked _____

Signature _____

Stub 1

Date _____, 20 _____

Paid to: _____

For: _____

Previous balance _____
Amount of check _____
Deposit _____
New balance _____

Check 2

Contractor I.D. # _____

Check No. _____

Date _____, 20 _____

Paid to the Order of _____ $ _____

_____ points credit for construction class.

Bank
TechEd   Total minutes worked _____

Signature _____

Stub 2

Date _____, 20 _____

Paid to: _____

For: _____

Previous balance _____
Amount of check _____
Deposit _____
New balance _____

Check 3

Contractor I.D. # _____

Check No. _____

Date _____, 20 _____

Paid to the Order of _____ $ _____

_____ points credit for construction class.

Bank
TechEd   Total minutes worked _____

Signature _____

Stub 3

Date _____, 20 _____

Paid to: _____

For: _____

Previous balance _____
Amount of check _____
Deposit _____
New balance _____

Copyright by The Goodheart-Willcox Co., Inc.

Construction Company Simulation

Check Stub 1

Date _____, 20 _____
Paid to: _____
For: _____

Previous balance _____
Amount of check _____
Deposit _____
New balance _____

Check 1

Check No. _____
Contractor I.D. # _____

Date _____, 20 _____

Paid to the Order of _____ $ _____

_____ points credit for construction class.

Bank
TechEd   Total minutes worked _____

Signature _____

Check Stub 2

Date _____, 20 _____
Paid to: _____
For: _____

Previous balance _____
Amount of check _____
Deposit _____
New balance _____

Check 2

Check No. _____
Contractor I.D. # _____

Date _____, 20 _____

Paid to the Order of _____ $ _____

_____ points credit for construction class.

Bank
TechEd   Total minutes worked _____

Signature _____

Check Stub 3

Date _____, 20 _____
Paid to: _____
For: _____

Previous balance _____
Amount of check _____
Deposit _____
New balance _____

Check 3

Check No. _____
Contractor I.D. # _____

Date _____, 20 _____

Paid to the Order of _____ $ _____

_____ points credit for construction class.

Bank
TechEd   Total minutes worked _____

Signature _____

Individual Earnings Summary

Name _____ Student I.D. # _____ Class Period _____

Emp #	Job Cat	Dates													Total Min	Total Hours	Rate ($/Hr)	Total $

Summary $
Management total (Job Cat x .1)   _____
Labor total (All Job Cat's x .2's)   _____
Special total (Job Cat 25-30)   _____
Contract Profit (+) or Loss (−)   _____
                                       Total _____

Emp # = Employer number
Job Cat = Job category

NOTES:
Management total = Sum of all times from Job Cats that end in .1
Labor Total = Sum of times for all Job Cats ending in .2

Copyright by The Goodheart-Willcox Co., Inc.

Contract Summary

Contractor _____
Contractor I.D. # _____

Module _____
Contract _____

Class Period _____

Stu ID #	Date and Time (minutes)																											Total Min	Total Hours	Rate $/Hr	Total $
CTM																															
																														Total $	

SUMMARY
Mat. Total $ _____
Mgn. Total $ _____
Labor Total $ _____
Grand Total $ _____

Contractors ID # _____
Module _____

DTL																															
CTL																															
Dtot																															
Ctot																															

MATERIALS PURCHASED
PO # $

Total: $

LEGEND:
DTL = Daily Total Labor
CTL = Cumulative Total Labor
CTM = Cumulative Total Management
StuID # = Student ID #
DTot = Daily Total Minutes
CTot = Cumulative Total Minutes
$ = Grand Total from PO
PO $ = Purchase Order #
Mat. Total = Material Total
Mgn. Total = Management Total